LAND SURVEY DESCRIPTIONS

By the late
WILLIAM C. WATTLES
Former Chief Engineer
TITLE INSURANCE AND TRUST COMPANY
LS 1071, Cal.
RE 1193, Cal.

Revised by the late
GURDON H. WATTLES
LS 2260, Cal.
RLS 1080, Nev.
Consulting Title Engineer &
Assistant Vice President, Retired from
TITLE INSURANCE AND TRUST COMPANY &
PIONEER NATIONAL TITLE INSURANCE CO.

Published by:
Wattles Publications
P.O. Box 1706
Tustin, California 92681-1706

10th Edition
9876

Library of Congress
No. 74-78750

Printed in the U.S.A.
by
Parker Graphics

CONTENTS

[For INDEX see pages 139 and 140.]

PREFACE TO THE TENTH EDITION

Over the years this book has been used as a reference to words, forms and interpretations and as a guide to methods of delineating land boundaries through the written word.

Many individuals have expanded their description writing through the application of the principles propounded herein.

Fundamentals do not change but time and custom exert an influence upon their interpretation and use.

Laws are changed, supplemented and in some cases deleted.

Court cases affect the interpretation of laws and consequently the application of the principles involved.

Questions have been asked concerning some of these matters.

With this in mind, the contents have been carefully reviewed. Some explanations have been revised or expanded for clarification. References to codes have been updated.

Every care has been taken to make this a publication of maximum value.

Gurdon H. Wattles

1881 IN MEMORIAM 1967

This tenth edition is a living memorial to a great man, a great teacher, Dean of Surveyors, my father, William C. Wattles. May the details and wisdom which he has imparted through these pages continue to enlighten and help those who tread this avenue of service to their fellow man.

Gurdon H. Wattles

GENERAL FEATURES

Every conveyance or document relating to the transfer, incumbrance or location of land, requires a description of the area involved for purpose of identification and permanent record. The need for precision of statement in such documents is self-evident. 100

The division of land into particular areas by survey and plat, is the foundation from which descriptions are derived. The first evidences of such division are the monuments on the ground and the record of the plat; descriptions written from these evidences were primarily of the metes and bounds, or boundary recital type, and generally covered acreage parcels of greater or less extent. Occasionally an area was divided into smaller parcels, town lots, a map filed, and the lots described as such by reference to the plat. Subsequent divisions of the larger parcels were made, either by the method of survey and description therefrom, or by description with subsequent survey. The difficulty of boundary definition and monument reference of the early days has been largely overcome, due to the almost universal division of land in this country, either by the government system of sectional surveys, or by areas within grants and ranchos, and their subsequent resubdivision into smaller parcels; recorded plats of such subdivisions showing boundaries and related features have been an additional aid. 101

While the metes and bounds form of description is always applicable to any parcel, if sufficient information is available, and is usually preferable in large, disjoined areas or very irregular parcels, the sectional system and the method of platting many parcels together in a subdivision, present a valuable means of defining boundaries with a minimum of language and a maximum of information relating thereto. This condensed form has been especially developed in California through the medium of legislation known as the Subdivision Map Law. This law provides for the recording of maps of an aggregate of parcels, defining the procedure therefor and specifying the information to be shown thereon. Because of the amount and value of that information, a simple description, properly referring to a parcel of a recorded map, covers and conveys all the data which could be furnished by an involved metes and bounds description, and more, as well as reducing the hazard of clerical error.

The written record of land holding embraces the description of it in terms which can be interpreted to the extent necessary to identify it, both theoretically and physically. 102

Written description is to the record as the possession is to the ground. This includes boundaries, areas, monuments, and claims of title of adjoiners. The title of land depends on the recorded evidence and the physical occupation; each should be consistent with the other to assure good title.

Reference to portions or all of maps or to the book and page of recorded documents may be substituted for descriptions under certain circumstances.

Some type of description is necessary in every instrument dealing with land areas where a location of the same is a factor in the terms of the instrument.

Any locatable description of land must have a base of location, either by reference to geographical coordinates, or lines of Government Township Survey, or Rancho boundaries or corners, or recorded subdivision, or to adjoining lands of record, or to physical monuments which themselves are relatable to points of record, or to any point or line which can be identified as to its relation with other record. Unless such reference is had, the description, as a matter of record, is void for want of certainty, or may be considered valid only as between the parties to the instrument on assumption that the language of the description discloses to said parties the actual intent thereof.

103 In the beginning the primary reference for land location was point position by latitude and longitude. This is usable in any area, especially that in which other types of land division have not been established.

The U. S. Government has surveyed and platted large areas of public land by the system of rectangular survey which is based on the location of areas known as sections, approximately one mile square, grouped in townships approximately six miles square, which are designated by successive numbers Townships North or South and in Ranges East or West of an established Base (parallel of latitude) and Meridian (meridian or longitude) the intersection or coordinate point thereof arbitrarily selected, having a known latitude and longitude.

Lands patented to individuals by the Federal Government under territorial acquisition and treaty, such as California ranchos, pueblo land grants, etc., and by the State for tide land locations, swamp and overflow lands, etc., for the most part have been surveyed and the maps filed for such patents.

These areas have been further divided by separate survey, and maps or descriptions, or both, of these smaller parcels appear in the record.

104 Fundamentally, any owner of land may subdivide the same in any manner he sees fit, and convey any part of it. However, certain regulations have been deemed necessary by the various governing bodies, and there are certain specified methods and restrictions imposed on the land owner.

The U. S. Government established a system of dividing the public lands, set out in the "Manual of Instructions for the Survey of the Public Lands of the United States." This includes the method of laying out mineral locations and claims, and deals with other types of land classification such as tide lands, swamp and overflow, etc., applicable to disposal of land by the Government to individuals or states. The various states, by statute, have supplemented this system to care for specific needs of state disposal and control.

The rancho surveys and patents were executed by authority and under instructions of the Government.

Certain tide land and overflow land surveys were executed under 104
authority of the State.

In California, subdivision of land into tracts, with lots, streets and other areas has been controlled by means of Subdivision Laws, passed, modified, and changed from time to time by various Legislatures. Refer to the Map Law in effect at the time desired, in the Statutes of the State; also refer to the Codes and General Laws of the state for special matters concerning the division of land.

The same procedure must be followed in any other state, viz., the investigation and conforming with the codes and laws of the state under consideration.

In California the making of surveys and filing of maps thereof together with the necessary engineering in connection therewith are controlled by the provisions of the Licensed Surveyors Act and the Civil Engineers Act, both of which are given in sections of the Business and Professions Code under Chapter 15 Division III and Chapter 7 Division III respectively. No person may practice surveying or engineering as a Principal unless he is licensed or registered under the provisions of the Acts, or specifically exempted therefrom.

DESCRIPTIONS

200 Descriptions for documentary and record purposes are invariably predicated on data derived from some previous survey or record.

The most satisfactory description from a title standpoint is one in which a minimum of bearings or distances is stated, using only those necessary to determine line of boundary or ties thereto which are not otherwise locatable by recitals to adjoiners of record; since dimensions give way to recitals of record and are therefore "more or less" in character, they are more informational than factual, and their use in actual relocation is to be discouraged unless verified by proper analysis. It is proper, however, for identification, to give such dimensions as will aid in interpretation of the description.

A good title description can be drawn with a minimum of information but a correct survey location of a good title description is not always possible. In such a case, the survey boundaries may be made in such a manner as to be safe without conflict and the uncertain area be left out. The proper fixing of doubtful or uncertain title boundaries is established by agreement or court proceedings.

Every description must recite one or more ties to monuments of record, to identify and locate the described land.

Monuments are physical or legal; physical may become legal when recited in conveyance or shown on instruments of record or so declared by a court decision; legal monuments in an instrument of record are any named boundaries or parts thereof whose positions are fixed or determined by another instrument of record or are created by the instant instrument itself. The identity between physical and legal monuments may have to be determined by the court.

201 *Essential features of Descriptions*

1. Precision of statement of items involved.
2. Complete definition of all terms.
3. Susceptibility of proper legal and engineering interpretation.
4. Freedom from conflict in recitals.
5. Freedom from conflict with descriptions of adjoining areas.
6. Definite showing of intent.
7. Mathematical and physical accuracy of dimensions.
8. Physical locatability through accurate analysis of its terms.
9. Clarity of expression.

10. Freedom from excess verbiage and redundancy.
11. Brevity without uncertainty or ambiguity.
12. Symmetry.

The above may be condensed to: Accuracy, recitals and ties, continuity, using these terms in their fullest meaning.

Comment 202

1. and 2. are necessary to prevent uncertainty and disclose intent.
3. Legal interpretation frequently differs from engineering analysis. Descriptions should reconcile both viewpoints.
4. Double calls or inconsistent statements produce conflict.
5. Specifically cured by proper recitals of adjoining boundaries.
6. The description is the written evidence of the physical position and its terms should specifically identify such position.
7. Precision is necessary in the record so that it may be duplicated in the field.
8. Should be so bounded and tied that monuments recited in the instrument may be identified with those on the ground. It is not necessary for such identification, however, to use bearings and/or distances between or along such monuments, although sometimes advisable for clarity.
9, 10, and 11. Loose language, immaterial and superfluous statements, words or phrases which add nothing to the meaning, should be eliminated, provided the precision, accuracy and intent are not disturbed thereby. Successive repetition of the same item when there is no different item of the same kind (such as "to the NE corner of Lot 7; thence along the E line of Lot 7 to the SE corner of Lot 7," when Lot 7 is the only lot in question) is too replete.
12. The artistic, neat, and compact description is more readable, understandable, and efficient. There is also a beneficial psychological effect.

Descriptions are composed of caption, body and qualifying clauses **203**
such as exceptions and reservations which clauses may be written in any order of sequence.

Descriptions may be divided into two or more parts or parcels for convenience in form or for segregation with respect to other considerations in the document. Separation into fee and easement parcels is advisable. Any parcel in a description may be referred wholly or in

203 part, by recital of lines or points, or other manner, to any other parcel in the same description.

Every description should carry recitals of

1. The city (if any), county and state in which the land is situated.
2. A statement or recital of the larger area or areas out of which the instant land is carved, together with a correct reference to the record of such larger area.
3. A Point of beginning, if the description is in the metes and bounds form.
4. The closing statement "to the point of beginning" if the description is in the metes and bounds form.

Where bearings are used, there should be sufficient evidence by statement, recital of record line with its bearing, or by other means, to determine the base (or meridian) of bearings used in the description. Unless qualified, or disclosed by intent of the instrument, bearings are to be assumed from an astronomical base determined by analysis of the description as a whole (see the section on "Lines" herein.)

A description should be accurate, specific and concise in all its terms. The exception to this rule lies in cases when it is necessary to establish title to an uncertain or indeterminate area by fixing a definite boundary with an adjoining title; in such case, *only* the definite boundary is accurately defined, the remainder of the area being described in such a way as to include whatever there may be in the old title.

Never omit the directional call of the tie distance even if obvious, such as "beginning at a point in the Northerly line of Lot_______ distant *Easterly* 27 feet from the Northwest corner (or most Northerly corner)"; a case may easily arise wherein the Northerly line may be a broken line, and the point of beginning could be interpreted equally well as lying either SEly (Ely) or SWly (Wly) from the recited corner if "Easterly" were omitted, thereby producing uncertainty.

If the point of beginning is not at a record point or corner, recite the beginning point as at the certain course from the corner, with appropriate ties, provided only one course is used; if more than one course is required, then begin at the reference corner, and successively recite the required courses therefrom to the "true point of beginning"; then carry out the description of the property and make the last course thereof recite "to the true point of beginning."

Some writers prefer a form of distinguishment such as "Beginning at . . . thence . . . to the point of commencement; thence . . . (follow the description of the property in question) . . . to the point of commencement." 203

Do not place figures representing difference of quality or quantity in juxtaposition; separate them by a phrase to prevent misinterpretation.

EXAMPLE: "thence Sly along the E line of said Lot 2, 15 feet."

This can be construed under some circumstances as Lot 2 a distance of 15 feet, or as, "said Lot" (previously noted as 2) a distance of 215 feet. The proper form is "Lot 2 a distance of 15 feet" or "thence Sly 15 feet along the E line of said Lot 2."

Every line of boundary in a description must, in one form of recital or another, state its identity with the corresponding common boundary of the adjoiner title (if such adjoiner exists), if conflicts are to be avoided, except in the following cases:

1. if the mathematical conditions and position of the adjoining common boundaries are precisely conformable and/or the description direction of courses of the instant property follows exactly in direction and dimension, that of the adjoiner;
2. if the recitals of monuments, physical or legal, are identical in the adjoining land descriptions;
3. if the line or lines in question are free, that is, original projections into the larger area out of which the instant property is taken, and which are not boundaries of any previously described area;
4. if the described line is being established as a substitute or agreement line for uncertain or unlocatable boundary.

If more than one reference, or recital, or tie, is made for a given point or line, the first is normally the superior, in case of conflict or uncertainty in the ties; however, this may be rebutted by the majority conformity of other ties, or by expressed intent in the instrument, or by status of title. If it is desired to express definitely the superiority of one tie over another, appropriate language should be used, such as ". . . to the NE corner of Lot 6, being also the SW corner of the land described in deed . . ." where the words "being also" indicate a secondary tie, subordinate to the directly stated one. However, if the following expression is used ". . . to a point in the East line of Lot 10, being 200 feet Northerly from the SE corner and also 175 feet Sly from the NE corner of said Lot . . ." then the ties should be assumed of equal weight, and the point located proportionately to the noted distances between the recited corners.

With certain exceptions, disclosed by the form of description, or 204
the expressed intent of the instrument or the superiority of pre-

204 cedence of calls, the following is the normal order of superiority of various types of call or recital, or monuments:

1. Monuments, physical or record, recited in the description.
2. Monuments, physical or record, not recited, but disclosed by analysis of the record and correlated information. Normally, natural monuments (trees, rocks in place, streams, etc.) will have superiority over artificial (stakes, pipes, etc.) but the contrary may be shown.
3. Recital of boundary of record or definite relation thereto as the line of traverse or boundary of instant property.
4. Recited distance and/or bearing ties to record corners or lines not on boundary of the instant property, unless such recitals are definitely secondary or informative.
5. Distances on boundary.
6. Bearings on boundary, or angles between lines.
7. Areas, stated as a corollary or addendum to the description.

   Note that a description given as a specific area (*e.g.*, "the E 6 acres of Lot_____") converts the "area" status to a "dimensional" status and falls under #5, as a variant thereof.

A physical monument not recited in the description, but identified as set prior thereto, and on which the description is based, and which is substantially conformable with the dimensions in the description and known to be in original position, will hold as physical evidence of title conveyed by the description, as preferable to the recited record monument in the description.

A physical monument known to be set subsequent to the date of the description, is not acceptable as superior to the recited record monument, and not safe to use as representing such record unless its position is verified by conformity with the record, even though it may become a proper title corner by any of several means, such as consent, estoppel, agreement, court decision, etc. In the event such monument becomes established as a title corner, then the description properly should be amended to tie specifically to such monument, and new deeds recorded to correct the record.

In general, physical monuments are superior to record monuments if set prior to and intended to determine such record; the record is superior to the physical if the physical is set subsequently, although purporting to identify the record, particularly if the analysis of records shows uncertainty or variable position of location.

Bearing or angle from next preceding course will hold against next distance subsequent to a course if that distance be to a called monument but not recited specifically as a tie; for example: "(1) N 10° E 400 ft.; (2) thence N. 70° W 350 ft. to the east line of Lot 6; (3) thence along said east line S 4° E 270 ft. to the SE corner of said lot..."; the recorded angle of 80° between courses (1) and (2), will hold as against an actual measured angle of 79° 40′, turned from the north extremity of course (1) to a point measured N 4° W 270 ft. from the SE corner of Lot. 6. The contrary may be shown under certain circumstances of analysis. 204

A specific call of area in a description (such as ". . . containing exactly 5 acres . . .") with one or more doubtful boundary calls or dimensions, may define the boundaries with precision, regardless of normal superiority of other calls. On the other hand, the "exact" area call may become in effect "more or less" if the description is specific, complete, and particular with proper boundary recitals and ties. The recital of area in a description may be good evidence to discover position of an error in boundary or closure, on the assumption that that adjustment will be most probable which will most nearly approximate the recited area.

Any discovered error, either from field location, or mathematical consideration of the description, properly adjusted or balanced, or specifically determined as falling in a certain place in the description, whether it be in the boundary of the land or in a tie to the boundary or an incorrect call to a monument, will tend to change or differentiate the normal succession or superiority as noted above.

In the absence of any information or acceptable analysis showing the actual or probable location of an error in closure, such error is thrown into the last course of boundary which recites "thence... to the point of beginning." Except that in the case of internal ties to boundaries, monuments, etc., the bearing and distance may be variable.

It is never competent to locate (for title purposes) courses of free boundary (*i.e.*, boundary without record adjoiners) by proration of traverse to mathematical closure, except:

1. when a known and uniform variance between the original tape measurements and the later tape measurements exists;
2. when the court decrees the proration as a proper means of adjustment.

205 *Bearings* (See also section on "Lines" herein)

Angular quantity in a bearing is a directional dimension, corresponding to the length of a line in extent of dimension.

Every description must show within itself the base of bearings used, either stated or indicated by reference to a line, or plat of record, or the presumption that it is referred to some specific meridian bearing North, determined astronomically. The presumption is generally modified to refer to the meridian through the point of beginning of the description or to the meridian through some specifically designated point of boundary. It is presumed, unless otherwise stated, that all bearings in any description are to same base.

If no base is stated or implied by reference, the normal presumption is that the base is the same as that of the subdivision out of which the described area is taken.

Where necessary in location, astronomical observation must be made; usual objects are the sun, by direct observation or by solar attachment, and Polaris at elongation; any celestial body may be used. Polaris at elongation is the simplest and most accurate method; with care, a direct solar consisting of the averages of several readings, direct and reversed, can be determined within 15 seconds of Polaris observation.

Relation between bases on different descriptions having a common line is shown by the differing bearings on that line; further verification may be made if there are two or more lines in common, or on prolongations or parallel, by identity of angles in each description between such lines. A recital of the different bearings, with their source, on a common line, is often a good means of identification of the line and the variance in base of bearing.

Sense of direction in bearing is essential in descriptions although not material on map. N 10° E and S 10° W have same angular variation from meridian but opposite sense of direction.

"North" may represent an actual bearing, or a directional position, but "northerly" is purely directional, except that legal construction of "northerly," when not qualified by the context, is taken to be astronomical north or "North" bearing on the base of bearings of the description. Directional position is preferably referred to by the direction with suffix "ly," such as "NEly" instead of "NE." Any directional term not qualified by a tie, must be construed as exact; (*e.g.*, "NEly" would become N 45° E). Dimensional bearings (*e.g.*, S 37° 20′ E) never carry the directional "ly" (Sly 37° 20′ Ely).

Mathematically, the limiting angles for direction of north or northerly are 22½° E or W of meridian; limiting angles for east or easterly are 22½° N or S of east; between 22½° E of N and 22½° N of E, direction is NEly; however, common practice and custom allow variations of as great as 37° from N or E, to be recited as northerly and easterly. It is dangerous to overstep the above limits to any great degree.

In certain localities, custom and use assume and call the direction of a prominent feature such as a highway, ocean, river, etc., "North," or "West," when the actual bearing of the feature is very nearly N 45° W; this use, although error in fact, must be considered in analysis of description, especially in the directional part lot form.

Azimuth

In general, azimuth is the direction of a line stated in terms of its angle, measured clockwise, from a specified origin (meridian) line.

The origin line is usually an astronomical meridian. In the United States, the general practice in land surveys is the use of a "South" meridian. Army, Navy and Air Force use "North" meridian. Canada uses a "North" meridian.

In a description, azimuth takes the place of bearings and may be stated thus: "thence on azimuth 38° 20′ a distance of 625 ft. . . ."; the recited angle showing both the variation from meridian and sense of direction from initial point of the course; in the above instance, the line is the same as one bearing S 38° 20′ W. Azimuth of 280° has bearing S 80° E.; azimuth of 218° 20′ has bearing N 38° 20′ E.

It is essential in description work that the azimuth origin (S or N) and the position of the azimuth meridian used as a base be stated in the description.

Use of azimuth in descriptions is not common except when used in the coordinate system; some map work shows azimuth directions, and much computation uses azimuth angles instead of bearings.

Angles

Because of the common practice of measuring angles in the field by repetition, a large part of the field work and notes, and much of the map work shows direction of lines by angles between them.

Descriptions may be written using recited angles or deflections between courses instead of bearings.

Although generally converted into bearings for description work, occasionally the angles are used. A common form is as follows: "thence Wly on a line deflecting Nly 10° 16′ from the prolongation of the last above described course, a distance of 346 ft. . . ."; or "thence SEly 275 ft. along a line making an angle of 147° 52′ toward the SW, with the last described course. . . ."

206 *Distances*

The distance between two defined points is governed by the *actual* distance between them regardless of the recited distance in the description, and such recital is, in effect, "more or less"; however, if only the initial terminus of the line is fixed, the other terminus is determined by the recited length, which then becomes exact.

Distances given in a description or shown on a map are "horizontal"; *i.e.*, level, or perpendicular to a plumb line, as distinguished from "slope" distance. A horizontal distance is in fact, a measurement along the arc of a curve, the center of the circle of said curve being the center of the earth.

Unless otherwise qualified, a stated distance is taken on a *straight line* between the termini; the terms "straight line" or "direct line" are really unnecessary but frequently used to identify, or make clear, the relation between the instant line and others in the same description; these terms are commonly used in descriptions of lines dividing lots wherein the lines are described by no other means than references to their record extremities.

Stated distances along the same or adjoining courses, if properly given and described by the necessary references, become associated in a proratable relation, and vary proportionately with each other, but the evidence permitting such proration must be clear and definite.

EXAMPLE: The lot frontages in a block are proratable to the difference in distance between the block corners as recorded and measured; the dimensions of various parcels in a partition are proratable to the differences between recorded and measured boundary dimensions or areas.

If the language of the description gives superiority or definiteness of one distance over another associated with it, then the superior distance is fixed and the subordinate length must change to conform.

The recitals and terms in a description determine the exactitude or proratable condition affecting these distances; if the distance is recited between fixed monuments, it is alterable and subordinate to the actually discovered distance; if the recited line is free on one end, the called distance is exact; if the free end is also intended (but not recited) as the end of a free line coming from the opposite direction, the total of the two given distances being found to be *greater* than the actual ground (or record) distance, then the common free end *must* be established at the recited distance given for the line stated in the senior (earlier) description, the other line taking the remainder; a prorate of the distances is *not* justified from a title standpoint; if the total distance is less than the actual ground distance, then each segment takes its full amount and a gap is left between the free ends. It is understood that the direction (bearing) of both lines must be held by the senior described line in the first case but for each line in the second case; the bearing of the junior line in the first case is subject to analysis as to any change from record. **206**

If a deed calls for a lot on a map, the dimensions of the lot are locatable on a prorate basis with the balance of the lots in the block, but *definitely restricted* as to the manner of prorate by *all* the items of analysis applicable to the relocation of the block and lot lines by full survey, and also limited by the existence of acceptable lot corners at points between the extremities of the block.

Although it is always proper to include a prorate theory in any location analysis, the general rule is that proration is the last rather than the first consideration of method in fixing points of title on relocation.

Many descriptions, particularly of strip parcels (rights of way, roads, etc.), give distances by the "Station" method, based on some point as 0+00; the distance on any course is the difference of the station numbers at each end of the course. Often a change of alignment of part of the strip is made, and to maintain the stationing system, an equation is inserted at the point of return to the old line; likewise, independent stationing on two or more intersecting lines is often equated at the intersection. On maps the relation of direction and reference lines of the stationing is usually apparent, but in description work it is often vague or uncertain. *Any* equation recitals in description should carry specific reference to the direction and line for *each* station given; *e.g.*, . . . thence N 46° E to station 76+42.26, equals (or, being the same as) station 78+39.00 going easterly on realignment of said center line, and is also the intersection with the center line of R Street at station 42+60 of said center line of R Street going South. . . .

Curves **207**

Except in special cases or in certain railroad right of ways (mainly in track construction), all curves in title and land surveying are plane circular curves; they may be combined successively with straight lines (which may or may not be tangent to the curves) or

207 with one another as reverse or compound curves, or as intersecting curves.

Mathematically, a plane circular curve is fully determined by *any* two elements or dimensions; those most frequently used are radius, length, tangent (semi tangent) and delta (intersection, or central angle); in description work, the most common are radius and length, with the central angle often added as a check or to make calculation more easy.

In description work, at least five elements are necessary to determine a curve:

1 and 2. The elements of dimension (R, L, angle, etc.).

3. The direction of curvature (concave to SE; convex to W—convex rarely used).

4. The direction of extension (travel) along the curve (Ely; Sly).

5. The relation of the curve to the next preceding course or curve if not tangent and the relation of a following curve as to whether it is compound, reverse or at a point of cusp.

Optional additions to the five elements may include one or more dimension elements (central angle or angle of arc, tangent, chord bearing and/or length, etc.); relation of the curve to the next succeeding course or curve; any other data for clarity or interpretation. It is evident that the additions must be precisely consistent with the necessary elements given.

In reciting such additions, name them in the order of intended superiority so that any inconsistency may be properly analyzed; for example: if length is recited prior to central angle even though calculated therefrom, it will probably prevail over the angle, while if the angle is first stated, the weight of probability tends toward the angle.

Length of curve is the most widely used of all curve elements. It is invariably a computed dimension if dependent on survey data; or if the length is assumed, then the survey can supply only one dimensional element (usually an angle or radius), all others being computed. A curve cannot be directly located, but is dependent on locatable points on the arc.

The following description includes the various forms of relation between curves and straight lines; the portions in parenthesis are optional additions; the form is one of several which may be used, but is probably the simplest. See Fig. 2.

N 50° E 726 feet to the beginning of a curve* concave to the S having a radius of 400 feet; thence easterly 300 feet along said curve

*The relationship of line to curve, curve to curve, curve to line is tangent unless otherwise stated.

(through an angle of 42° 58′) to the beginning of a reverse curve concave to the NW having a radius of 326 feet (through which a radial line bears S 2° 58′ W); thence easterly 275 feet along said curve to the beginning of a compound curve concave to the W having a radius of 100 feet; thence northerly 100 feet along said curve; thence N 12° 40′ W 173 feet to a point of cusp with a tangent curve concave to the NE having a radius of 200 feet (to which a radial line bears S 77° 20′ W); thence southeasterly 200.01 feet along said curve (through an angle of 57° 18′) to a non-tangent line (a radial line to the end of said curve bears S 20° 02′ W); thence S 60° E 150 feet to the beginning of a non-tangent curve concave to the S having a radius of 370 feet to which point of beginning a radial line bears N 1° W; thence easterly 193.73 feet along said curve through an angle of 30° 00′ to the beginning of a non-tangent curve concave to the S having a radius of 420 feet to which point of beginning a radial line bears N 10° W; thence easterly 250 feet along said curve.

An area within the full circumference of a circle may be described in form as follows:

That parcel of land included within the circumference of a circle having a radius of 78 feet; the center of said circle being the NE corner of Lot 12.

To make certain the intent of tangency of a described curve with a record line or curve subsequent, the following form may be used:

N 50° E 273 ft to the beginning of a tangent curve concave to the NW, having a radius of 30 ft, which curve is also tangent to the (curved or tangent) SWly line of A Street; thence NEly along said curve 46.22 ft to said SWly line (of A Street).

The parentheses are optional.

This form places the more or less distances outside the curve location or radius and demands a tangency.

The concavity direction of a curve is that of the direction of the center of the circle of said curve from the mid-point of the arc described; a described segment of a record curve may be concave to the East, while the whole of the record curve may be concave to the North. Fig. 5.

Direction of travel along a curve may be recited in multiple if the central angle is near to or greater than 180°, as—thence Sly, Ely and Nly along a curve concave to the north.

The center of a curve is the midpoint of the described arc, as distinguished from the *center of the circle* of which the curve is an

207 arc of the circumference; a radial bearing to or from the "center of said curve" is a bearing of the radial line passing through the midpoint of the arc. Fig. 5.

It is frequent practice to recite "center of curve," when "center of circle of which the curve is an arc," is intended; such form should not be used as it requires interpretation, and sometimes the intent is not clear.

Although the bearing of a radial line on a map may be shown in either direction (N 10° E, S 10° W), the description should always recite the proper direction conformable with the context; the extension of a radial line is correctly *from* the center of the circle *to* the circumference. For example, use: "a point in a curve concave to the SW having a radius of 200 ft *at* (*to, through*) which a radial line (or, a radius) of said curve bears N 37° E," or—"a point in a curve concave to the SW having a radius of 200 ft *from* which a radial line of said curve bears S 37° W (or, from which the center of the circle of said curve bears S 37° W).

The radius of a curve does not extend beyond the circumference of the circle, and any description of the continuation of such radius should recite ". . . thence along the prolongation of the radial line of said curve through said point. . . ."

Curves are tangent if they are compound or reverse; they have a common tangent line at right angles to the common radius at the point of compound or reverse.

A line "at right angles" or "normal" to a curve is a radial line of the curve, or its prolongation. The terms are frequently used but are not as satisfactory as "prolongation," or "radial line."

The expression "a radial bears N 30° W" to express the idea "a radial line of said curve through said point bears N 30° W" is too abbreviated for good usage and does not fully express the idea.

Use of "degree of curve" instead of radius, is permissible in description if specifically qualified as to derivation; the radius of a certain "degree" may vary according to the engineer who uses it (see Nagel, Searles, etc., handbooks).

Reference in description to "Taper No. 2" etc. to define particular forms of spiral curves is permissible only when the context indicates the special form of spiral intended to be used. Bear in mind that Railroad Company records are not records which are generally available and reference to their maps, or degree of curve, or spiral, is incomplete without additional evidence showing intent; this applies also to Railroad center line Station recitals unless qualified.

Reference to "located center line" whether on straight or curved lines is unsatisfactory without other qualification, because it depends entirely on the fact of survey and relocation. Frequently a right of way is acquired or intended to be acquired as a strip of uniform width with the tangent portions joined by simple curves; the track location is nearly always tapered as to the curves; result of "located center line" is to make side lines of the right of way follow corresponding parallel tapers on the curved portions.

If a distance is intended to be given along the arc of a curve, it should be so stated (Ely 100 feet along said curve), otherwise (a point in a curve distant 100 feet from the East extremity of said curve) the distance is taken on a straight line (chord) between the designated points.

Parallelism 208

Straight lines are parallel if they maintain a constant perpendicular distance between them; since lines have no limit as to length, they may be recited as parallel if the perpendicular distance between a recited segment of one and the prolongation of a recited segment of the other is constant; the statement of prolongation is unnecessary as it is presumed to follow; in some cases, however, it is advisable for clarity. Fig. 4.

Broken or multicourse lines are parallel when the corresponding courses are parallel and all courses are at the same perpendicular distance apart, respectively; however, if the several segments of such lines are respectively parallel but at variant distances apart as between different segments, the two lines as a whole cannot be considered parallel. Fig. 8.

Curves are parallel if they are arcs of concentric circles in the same plane; usually limited to those arcs lying between or nearly between the same radii from the common circle center; curves having the same length radius but a different circle center, even though the arcs lie in the same quadrant, are *not* parallel. Figures 6 & 7.

Arcs of 2 circles not in the same plane but in parallel planes are parallel if the centers of the circles of such arcs lie on a common axis perpendicular to such parallel planes; usually limited to those arcs lying between the same or nearly the same planes perpendicular to such first mentioned parallel planes which intersect on the common perpendicular axis.

Lines 209

1. A line in geometry is the path of a moving point or the intersection of two surfaces.

2. A line in geodetic survey may be straight, as a line of sight between observed stations, or curved, as a determined length and/or

209 direction on the earth's surface, the surface of the basic spheroid of reference, or lines related thereto; or straight as in the various forms of plane projection developed for cartographic and planimetric survey use.

3. A line in a land survey is a physical position defined by monuments or features indicating such position; it marks the division between areas, or the boundary of a parcel of land, or some correlated position (such as traverse line, tie line, topographic line, etc.)

4. A line on a map or plat is a mark, with appropriate notation of dimension or other quality, representing a position of boundary or correlated matter, as it exists, or is proposed to be created, through ground survey. Such a line will be referenced to other lines and features on the map by some form of dimensioning or by visual (scale) relationship.

5. A line in description or title is a statement in words of a dimensioned and directioned position, referenced directly or indirectly to a point or line already defined of record.

6. At least two elements are necessary to position a line, *viz.* direction (by bearing, azimuth, or angle) and length (units of measurement); if in the description the line is preceded by a curve a third element is necessary, the recital of relation to such curve (tangent, non-tangent).

Unless qualified, the following statements will apply to lines defined in 5 above.

A line may be straight, bent, jogged, or curved. Description may be given for a segment, or may refer to an aggregate of segments as a general line of boundary, between termini of such boundary.

Unless qualified, every line is presumed straight; such qualification may take any of several forms, such as: — — following the various courses of the boundary of land described in deed . . . , or — — — along the center of Black Creek — — , or easterly along said curve or — — — along said west line as follows: N 30° E 426 ft; N 27° W 200 ft — — —, or — — — thence in a general northerly direction along the easterly municipal boundary of the City of M — — . Although unnecessary, the form — — — thence northerly in a straight (or, direct) line to — — — , is often used to distinguish or emphasize the nature of the recited line as compared to a similar or associated line which may not be straight, or to connect two dissimilar points at considerable distance apart when the distance between them cannot be accurately stated, such as: — — to the NE corner of Tract H; thence northeasterly in a direct line, about 4000 feet to the intersection of the center lines of A and B Streets — — —. There is no question or need of statement of "direct" or "straight," if a line has a recited bearing (N 20° 42′ E), for the bearing itself defines straightness of direction in planimetric surveys, but the term is often used to distinguish or clarify the language.

A line, or boundary, is "free" if it is not limited or fixed by predetermined (record) positions of terminal or "in line" monuments or by ties to "off line" record monuments (irrespective of whether any of such monuments be legal or physical) or by a stranger ownership adjoining the bounded area. It may be said that the only technically "free" boundary is that partially or wholly surrounding an area which lies entirely within a larger area held under a single ownership. In this case the original description (and survey), except for the terminus of the tie line to a record point, may be considered free; but subsequently from such origin, the lines are bound.

No survey, map, description, title area, or boundary is competent or determinable for title purposes unless directly or indirectly referenced and related to some known point or boundary already existent or established by record. It follows then that every description must have a point of beginning tied or referenced to some previously established point of record; also that all parts of the boundary of an area (except free boundary) must properly be related to adjoining existing boundaries. These points of beginning and reference boundaries need not be specifically stated as such except in metes and bounds type of description; for example: the easterly 50 feet of Lot 6, Tract X — — , gives all required boundaries and ties, by reference to matters on map of Tract X and the 50 feet stated in the description; the need for a point of beginning is eliminated by consideration of the area and its boundary as completed units.

The only exception to the above is the case of non-existence of a record, such as unexplored or uncharted areas. In such event an arbitrary and physically locatable object is fixed as a base, or some reference point determined by observation, fixing latitude and longitude or both.

Every segment of a line (or the whole line if straight, or the full arc of a single radius curve) must have a dimension of length, actual, implied, determinable by calculation or survey, or have as termini points which can be definitely positioned. The dimension may be in any unit desired (feet, chains, varas, etc.) but, unless specifically stated otherwise, all dimensions in any particular description are presumed to be given in the same unit; it is rare, in practise, to find a stated length without specifying the unit name (such as — 460 —, instead of — 460 feet.) Dimensions on maps frequently are not

209 recited except in a general statement as to the scale of the map (1 inch = 100 ft) the unit stated being presumed to identify all distances known unless qualified in some particular.

Every segment of a line (or the whole line if straight, or the full arc of a single radius curve) must have a stated direction, extending away from its common terminus or junction point with the next previously recited portion of boundary (or tie line or point) in order that the description of boundary may proceed from its point of beginning, successively through the various courses, back to beginning. The specific application here is to the metes and bounds form of description, and to description of lines extending between fixed termini. In directional description forms (— — the E 50 ft — —, etc.) the stated direction may apply to a particular line or lines, or to the whole area in question, or both, and indicates directional position toward a fixed or record boundary.

Directions may be general (easterly, southwesterly, etc.) or specific (N 30° E, azimuth 258°, deflection 14° to the east, etc.)

General directions are effective only if there are other sufficient ties or references to fix definite position: for example: — — thence northeasterly 375 feet — — —, is indefinite unless the terminus of the line is defined, as, — — — 375 feet to the west corner of Lot 10 — — —; without such definite statement, the direction "northeasterly" would have to be presumed N 45° E, unless the line position can be fixed in direction by other matters developed from the description as a whole.

As the measurement of length is referenced by the stated unit of length, so is the specific direction of line position referenced by a base line, known as the meridian of reference. This meridian can be any line, arbitrarily chosen. Usually it is an approximate north-south line. Occasionally it is a true astronomic meridian (or parallel of longitude), but errors of observation or assumption generally destroy the precision of location and it becomes the approximate type. For certain special purposes in survey and computation checks, a line far from a north-south direction may be taken as meridian, have its extremities designated arbitrarily as "north" and "south," respectively, and be used in all respects, in the problem, as a meridian of reference.

Specific direction may be expressed (1) by directional angle measurement from the next preceding course of a description, or from any other designated line, (2) by azimuth, (3) by bearing from a reference meridian (astronomical or assumed), (4) by coordinates, (5) by combination of any of the above with or without other words of designation.

(1) Directional angles may be given from any line to the line in

question by stating the angle found between the lines (or their prolongations). It is necessary, in every instance, to state the position of the angles with reference to the lines, such as: — — thence easterly along a line making an angle of 105° 20′ toward the southeast, with the last above described line — —, or, — — thence deflecting to the east, 27° 15′ from the northeasterly prolongation of the last described course, — — or, — — thence northerly along a line perpendicular to the easterly prolongation of the south line of Tract M — —.

(2) Azimuth, as applied to land surveys and descriptions is the directional value of a line defined as the horizontal angle to such line, measured clockwise from astronomical south meridian through the initial terminus of the line. Azimuth may likewise be measured clockwise from astronomical north meridian. Conversion from South azimuth to North azimuth is made by adding 180° to south azimuth angles between 0° and 180° and subtracting 180° from south azimuth angles between 180° and 360°. Unless qualified, azimuths in planimetric surveys or descriptions must be given from a common (or single) meridian of reference.

Any description or map using azimuth directions must, for safety, state the zero origin either south or north.

The azimuth method and nomenclature may be used with, and from, any reference meridian whether north-south or other, if definitely specified.

In descriptions using azimuths for line direction, the specific recital should be made that the meridian of reference is a line (true S, or N, or other, as the case may be) passing through the (Nly, Sly, Ely, etc.) terminus of the (name or description) course of the described boundary; otherwise it might be assumed that each line carried its own true azimuth, independently of the rest.

(3) The most commonly used directional designation in land survey and description is the bearing.

A bearing is the horizontal (or plane) angle, between 0° and 90°, measured easterly or westerly from the northerly or the southerly projection of a meridian passed through the initial terminus of a line, to the line itself. Nomenclature states the meridian direction (N or S), the angle (20°; 47° 15′ etc.) and the direction of angular measurement (E or W); example: N 20° E; S 47° 15′ W; etc.

A description involving bearings must be interpreted on the assumption that each bearing therein bears its respective angular relation to the same or common meridian of reference, unless specifically stated or definitely identified otherwise; this assumption derives from the fact that land surveys (except those made for geo-

209 detic purposes over large areas) are planimetric, that is, made on the theory that the earth is basically flat, all measurements of length and angles being made in a horizontal plane (perpendicular to a plumb line, all plumb lines being presumed parallel), with the boundaries and areas developed and computed by the principles of plane geometry and trigonometry. Transformation of bearings to angles, or vice versa, under this plan, demands the single reference meridian, or a determinable equation, if more than one meridian.

Although the planimetric assumption is false as to the true shape of the earth, practically it is accurate for all ordinary limited areas of survey and description, the dimensional differences between the curve of the earth and the plane of the survey being well within the minimum allowable, or practical limit of ordinary observation; also, the planimetric method, used and accepted for designation, description and holding of land areas over hundreds of years, has become axiomatic as a mode of procedure.

The meridian of reference, commonly known as "base of bearings," is not often stated or directly used as such in a description; rather, there will be given the bearing of some reference or tie line, which is assumed, implied or designated as the base of bearings (*i.e.*, having a bearing which shows the correct angular relation to the meridian of reference); or a phrase (bearings from true meridian; bearings from base noted on Tract S; courses magnetic; etc.) may be added to the regular description. Since every description must have a base of bearings, the analyst must determine the same in each case according to the majority probability of the contributing factors.

In the event the description shows bearings but no evidence indicating the presence of a locatable meridian of reference, it will be competent to establish such meridian by astronomical observation. If such observation is had, the description should then be modified to identify the meridian by adding words in appropriate places such as: — — — N 30° E, bearings taken from astronomical meridian through the southerly extremity of last described course — — —, or — — — Beginning at — — — thence, with bearings from true meridian established through the monument at the southwest corner of Tract M herein described — — —; etc.

The cardinal directions, North, South, East, West, are bearings with the angular measurement from meridian omitted, since these angles are 0° or 90° (North is the same as N 0°; East is the same as N 90° E). In descriptions the word "due" is often added to the cardinal directions, apparently to give emphasis to the term as a bearing instead of a general direction, or call attention to the lack of angular variation. The presence of the word "due" has some-

times been considered synonymous with "true" or astronomical; this 209
concept is untenable since there is no significance or connotation permitting such interpretation (*e.g.*, "due" north is not "true" or "astronomical" north unless definitely so stated or indicated in the description). The word "due" in association with bearings is definitely superfluous and frequently misconstrued; it is not a satisfactory call and should not be used. If found in a description, it should be interpreted as though it meant "exactly" or "in a straight line" or "in the specific direction of," unless the context requires a different meaning.

The meridian of reference may be *any* line acceptable for the purpose of the description; the more common meridians are:

(a) Those used by descriptions or surveys of record adjoining or near the boundary in question;

(b) Meridian determined by astronomical observation on or near the boundary in question;

(c) Magnetic meridian ascertained by present-day compass or by difference between recorded compass declination and present observed declination at the same place, or by other means;

(d) Any line which may be convenient in a particular case to facilitate ground layout or computation, from which the bearing form of direction is used.

(e) The "grid" meridian determined by the development of a plane coordinate system for the area under consideration.

Comment

(a) Use of the bearing of a boundary line of an adjoining property (given in a description or shown on a map) as a base of bearings, is most convenient and useful for computation and comparison since it requires little or no equating of directions to reduce to necessary common base.

(b) If an astronomical meridian is intended as a base of reference and so indicated or implied in the description, use the form and qualifications stated in the above paragraphs concerning meridian to be used in descriptions with unidentifiable or unlocatable base.

It must be remembered that the *stated* astronomical meridian in the description, is the only line therein which is obtained or used geodetically; this line becomes planimetric as a base for angular measurements for all other lines in the description (except where definite statements are made to the contrary).

209 With such a meridian as with others, all "North" or "South" bearings are parallel with the reference meridian, and all "East" or "West" bearings are at right angles to the reference meridian.

(c) Certain localities, by custom, base the bearings of descriptions and surveys on the compass, or magnetic meridian. The practice is very inaccurate due to: (1) continuous and irregular diurnal and annual angular variation of the magnetic pole with relation to any fixed line on the ground; (2) inability to observe, accurately, such variations, or to predict any definite future relationship therewith; (3) the fact that the magnetic bearing of a line will vary throughout its length to a greater or less degree, depending on the direction of the line and/or on the local ground attraction of the compass needle.

It is customary in descriptions using magnetic bearings to make a statement such as: — — thence by magnetic bearings, the variation of the compass being 14° 30′ E — —, or similar phrasing, in the attempt to provide a second and more stable meridian (presumably true north but very seldom correct) from which the survey or description may later be retraced, after making a correction equation between the variation stated and that determined at the time of retracement.

The only satisfactory interpretation of a description using magnetic courses is to assume (1) that all given bearings are magnetic to a common reference non-magnetic meridian; (2) that such non-magnetic meridian is located as a straight line making an angle with the first course of the description composed of the algebraic sum of the recited needle variation angle and the bearing of said first course, if said course is definitely monumented on the ground, or with that course in the description, physically monumented, which most nearly conforms with the analyzed expressed or implied intent of the description.

Modern practice and accuracies indicate abandonment of the old form uncertain magnetic courses, and replacement with definite reference meridians by means of acceptable resurveys and correction deeds.

(d) Certain investigations as to source and quantity of error in a closed traverse or boundary are more easily examined by assuming an arbitrary base reference line, considering the same a meridian with conventional N,S,E, & W positions thereto, and converting the bearings of the given traverse to new bearings with corresponding angular relations to the assumed reference line as a meridian. After determination of the sought data, bearings and angles can be reconverted to the original meridian.

(e) If the angle points of boundary are positioned by coordinates, the direct bearings of the lines to the grid meridian are found by the ordinary trigonometric processes. (See 4 below)

(4) It is possible to determine both linear dimension and angular direction by the use of rectangular coordinates.

Coordinates of a point represent quantities, in the stated unit of length, which indicate distances, measured at right angles respectively from two reference lines or axes, themselves at right angles to each other. These axes are generally North-South and East-West lines, true meridian or nearly so.

The intersection of the two reference lines has coordinates o-o. Distance in the direction along or parallel with the N-S (approximate or true) reference line is designated as Y distance or Y coordinate; distance in the direction along or parallel with the E-W reference line (at right angles to the N-S reference) is designated as X distance or X coordinate.

The position of a point is given in description or survey as: — — point having coordinates X (or E) 20756.20 ft, Y (or N) 1642.75 ft — — (or similarly), meaning a point at the distances recited along or parallel with the X and Y lines (axes) respectively, from the o-o coordinate point.

The length and direction of a line between two points whose coordinates are given is readily ascertained by trigonometric calculation. If a bearing of the line is desired, it may be based on the angle with the Y axis, or any other line whose angle with the Y axis is known.

For convenience and freedom from use of negative values the o of coordinates is generally taken at sufficient distance West and South of the area of survey or description to permit the coordinates of any point in such area to be given plus values (*i.e.*, E and N). (Coordinate values are also designated as "X" for East and "Y" for North.)

In any description using coordinates, the base and direction of the axes (X,Y) should be stated or definitely implied, as for example: — — at a point having coordinates X 5962 ft, Y 4727 ft, based on the coordinate system of the C Water District; or — — — coordinates N 9999 ft, E 7000 ft, based on the Lambert Plane Coordinates, Zone 7, of the California Coordinate System — — —; etc.

If coordinates are used in calculation, any pair of axes may be used; the axes may be at any angle with each other and the o of coordinates (intersection of axes) need not be located outside of the area of operation; but the coordinate distances must be reckoned parallel with the axes and the algebraic sign of the values must be considered.

(5) In description involving any combination of the meridians of

209 reference in (1) to (4) above, there must be a definitely indicated or a stated relation at the point of change, between any two meridians, or the equation between the meridians may be stated for a given point.

The following example will illustrate relation and equation:

Beginning at a point in the northeast line of M Street distant along said line S 76° E 400 ft from the SE corner of Tract X; thence N 43° 40′ E 250 ft; thence northwesterly in a direct line to the most southerly corner of land of Black; thence along the SE line of said land N 37° 50′ E (record in Black deed) 527 ft to the most easterly corner of said land, said corner having coordinates of N 27565, E 52894, based on the Q Water District Coordinate System; thence S 43° 15′ E to the westerly line of Lot 15 Tract P; thence southerly along said westerly line to the northeast line of M St; thence N 76° W to the point of beginning.

Notes: (1) The bearing S 76° E on M St. is the base of bearings for the course N 43° 40′ E, giving an angle of 60° 20′ toward the east, between the two lines.

(2) The bearing N 37° 50′ E, by reason of the unstated bearing of the second course of boundary and also by the statement (record in Black deed), furnishes a new base of bearings for the following course. The relation between the first and second bases, in this example, must be determined by survey of ground location.

(3) The coordinates at the most easterly corner of land of Black will relate the bearings of the lines in the example with those in the description covering the boundary wherein the coordinates are given, providing a definite reference for comparison of title lines of adjoining properties.

(4) The bearings S 43° 15′ E is laid from the base noted as "N 37° 50′ E," making an angle of 81° 05′ toward the south, with such base.

(5) The lines showing no bearings require none, for description, since their extremities are definitely stated monuments. However, a properly executed survey of the description will provide bearings and/or angles between all lines of boundary, and the description May be rewritten, if desired, to a single base of bearings, provided the proper recital of ties and title calls is retained.

210 *Areas*

As in curves, areas are not directly locatable as such, but must be laid out by linear dimension boundaries, computed from the area data.

Unless the area is the controlling call in a description, there is no necessity for recital of it other than for indicative or informative purposes. Regardless of the recital of specific area (10 acres) or presumptive area (10 acres, more or less) it remains subordinate to the calls and dimensions of a correct description, or of a properly and correctly dimensioned parcel on a map. **210**

If a description (or a parcel on a map) carries a recital of area (even "more or less"), such area will prove valuable in showing intent or trend if there are irreconcilable errors or differences in the terms of the description, or in the physical conditions affecting ground location. Even though the given area is incompatible with the final location, the majority probability of correct location (as between two or more solutions) will be influenced by that most nearly conformable with the stated area.

For an area to certainly control over other recitals in a description, the area call must be specific and definite, overriding all other dimensions. Example: "The Easterly 10 acres, between parallel lines, of Lot 6"; or ". . . by a line parallel with the South line of Tract X, at such a distance as to enclose with the remainder of the boundaries exactly 3 acres"; etc. The courts may grant more latitude in area control factors if evidence and conditions warrant it.

If the area is free, that is, without record adjoiners, and the description is wholly metes and bounds, the recited dimensions must compute exactly, the required (exact) area dominating the description.

If the area is restricted in part of its boundary to existing record lines, there must be carried in the description at least one definitely variable dimension, to allow for a computed closure with the relocated record boundaries which will give the exact area called for.

If an exact area is called for in a description carrying physical monument recitals, there must be at least one portion of boundary which can be adjusted to reconcile the monuments and the acreage, since the physical monuments recited will control their existing positions.

Exact areas, although possible of record, may by reason of ground location become ineffective as against possession even though the possession was intended properly to cover the described area.

The following examples will show the effect of description wording on area consideration:

(1) ". . . 80 acres, being the E½ of the SE¼ of Sec. 15. . . ." is not exactly 80 acres, but

"The Ely 80 acres (which means between parallel lines) of the SE¼ of Sec. 15 . . ."

210 is a definite area, with the free (westerly) line specifically defined.

(2) ". . . one acre in the SE¼ of Sec. 23 . . ." is not locatable and the description is void for uncertainty, but

". . . one square acre in the SE corner of the SE¼ of Sec. 23 . . ." defines an exact acre in "square" (i.e. parallelogram) form having all sides equal length and the East and South lines being coincident with the East and South lines respectively, of the Section.

(3) ". . . the South (Southerly) 100 acres of the NE¼ of Sec. 19 . . ." is held to be 100 acres, exact, between parallel lines, the North line being parallel with the South line of the NE¼,

however, if the boundary of the property be irregular, the form "the Sly 5 acres of Tract M" is incomplete, being a doubt as to the direction or location of the free line; the free line should be described.

(4) ". . . 5 acres in the Northerly part of Lot 6, the Sly line of said 5 acres intersecting the East line of said Lot at a point 125 ft Sly from the NE corner of said Lot . . ." This allows the *direction* of the Sly line to be computed to include the proper area.

(5) ". . . the West 3 acres of Lot 10, the East line of said 3 acres having a bearing of S 16° 30′ E (based on the bearing of the North line of said Lot) . . ." This allows the *distances* on intersected lot lines to the extremities of the Ely line of the 3 acres, to be computed to include the acreage.

(6) It is essential, to prevent uncertainty, that the forms (4) and (5) above be used only in cases where the variable (free) line intersects two courses only of the larger boundary. The forms are most often used with quadrilateral shaped parcels.

(7) ". . . (metes and bounds); thence along said line S 63° E to a point from which a straight line to the point of beginning will include, with the remainder of the above described boundary, an area of exactly 4 acres . . ." This is similar to (4) or (5) above but allows more latitude in the shape of the figure from which the area is taken.

(8) The recital of fractional parts (E½, N¼, S½ of N 2/5ths, etc.) of a larger parcel, without qualification of the position of the free line or lines, is unacceptable in any description, because of the uncertainty of interpretation; this is especially true in fractional statements such as NW¼, SW⅓ of Ely½, etc.

Whole portion

Any parcel which may be denoted by number, letter or other designation on a record or quasi record map, or which may be referred to by reference to a recorded document in which the full description appears may be correctly and simply described by naming such lot by its number, letter, or record status together with the proper place of record. For example. "Lot 6, Tract No. 2746, in the City of Gardena, County of Los Angeles, State of California, as per map recorded in Book 10 Page 98 of Maps in the office of the Recorder of said County"; "Parcel 3 of Parcel Map in the city of Long Beach, county of Los Angeles, State of California as recorded in Book 26, Page 31, of Parcel Maps in the office of the County Recorder of said county."; or, "Parcel 14 of Licensed Surveyor's Map in the County of Los Angeles, State of California, as per map filed in Book 8 Page 37, Record of Surveys in the office of the Recorder of said County"; or, "That parcel of land described in deed to John Doe recorded in Book 7240 Page 37 of Deeds in the office of the County Recorder in the County of Los Angeles, State of California."

The last above form of description is especially useful in clearing uncertain or physically unlocatable interests of grantors or in re-describing difficult multi course boundaries previously fully set forth. It is also valuable in a description of a large parcel having numerous exceptions, each of which has been created by a deed describing the particular portion.

Lot or parcel division into parts 212

(1) Directional division, 2 parcels from 1.

(a) By recited dimension in designated direction (E 50 ft; S 10 Ft measured along W line; Nly 70 ft measured along the E & W lines) (See Fig. 1)

(b) By area, specified or proportionate (E half with line parallel with E boundary; S one third with line at right angles to west line; NW half with line set at midpoints measured along SW & NE lines of said lot)

(c) By exception of portion not conveyed
(Lot_____ except W 75 ft; Lot_____ except portion described in deed_____)

(d) By division line between parcels
(portion lying E of the following described line: ...)

212 Unless qualified (measured along the S line; measured parallel with the W line; etc.) the recited distance (E 50 ft of lot) denotes an area between parallel lines, at the recited distance apart, measured at *right angles* to the lot or parcel line adjoining the area. (Note: see definition and discussion of parallel lines.)

Care must be used in all the above forms of descriptions to avoid indeterminate conditions.

"Halves" and "Quarters" must be considered from two different aspects, each based upon its origin: Public Land Surveys or private land development.

The meaning of "half" and "quarter" as related to Public Lands both in title and ground location is defined in The Manual of Instructions for the Survey of Public Lands issued by the Bureau of Land Management, Department of the Interior (formerly General Land Office) of the United States. This is not based on area but rather on the breakdown of the section, quarter section, etc., according to the rules set forth in said Manual.

In private land development, "half" or other fraction denotes *area* unless qualified by a dimension recital.

The "E half" of a parcel has no definitely determined *W* line; ANY line dividing the parcel into equal areas can be construed as the line between halves, regardless of the shape of the lot. The intent of the deed description is not necessarily disclosed by the shape of the parcel on the map.

The only safe boundary description of "the E half," is "the W half" or "the lot except the E half."

There is no conflict in *title* between "West half" and "East half"; the question is in *ground* location when there is no statement of the position of the line between the halves.

Do not write descriptions on "the E half," without qualification, *unless* the other portion has already been described of record as "the W half," or "Lot_____ except the E half."

The line between "halves" of a lot is not a locatable line unless its dimensions with relation to the boundaries of the lot are given, such as—"the E half of lot, the W line thereof being parallel with the E line of said Lot"; or—"the W line thereof passing through the mid-point of the South line of said lot"; or—"the W line thereof bearing N 3° W"; or other specific ties. (See Fig. 12)

The common boundary of indeterminate halves of a lot should be established by a recorded agreement setting forth the proper ties of such boundary.

It is not competent to assume the limiting line of the half to be 212
parallel with the boundary line of the lot on the portion in which the half is situated. It is merely custom and assumption of intent to consider such line parallel; it may be otherwise, unless the description specifically qualifies it.

The above statements hold true for any division of a parcel into recited fractional areas (E one third; W one fifth of E half; etc.).

A recital of the W¼ of a lot may be specific as to its general position, if the lot extends generally Northerly and Southerly (i.e., the side or end lines bearing N or up to N 30° E or W) and construed as taking a strip across the whole edge of the lot; but a recital of the SW¼ may be indeterminate, especially if the bearing of the SW line is near the 45° direction, for SW¼ can be interpreted as a strip across the SW portion of the lot or as a SW portion of the lot formed by lines intersecting across the lot from N to S and E to W, such as the SW¼ of a Section. Be careful in the wording of such a description to give the intent without doubt. (See Fig. 24)

EXAMPLES (a) If it is desired to take a strip across the W portion of lot covering ¼ the lot use "the SWly (Wly) half of the SWly (Wly) half."

(b) If it is desired to take the ¼ formed by the intersecting mid lines, use "the SEly (Sly) half of the SWly (Wly) half."

Although the usual assumption of "N 10 acres" is an area of 10 acres between parallel lines, the S line being parallel with the N line, the contrary may be shown; as in "half," the acreage is an *area* and as such may be governed by prior boundaries.

If a considerable portion of the lot extends beyond the intersection of the described dividing line with the lot line, it is safer to use the following or a similar form for the dividing line; "a line extending (or–prolonged) from the mid point of the East line of said lot Westerly *through* a point in the Northwest line of said lot distant Southwesterly 70 feet from the Northwest corner of said lot." (See Fig. 18)

The common practice of describing "the North 270 feet" of a 300 foot lot, for the purpose of excluding the portion deeded for street is dangerous; a surplus in the lot is a frequent condition, and such description will leave the part of the lot described without access

212 to the street; use instead "Lot____ except that portion in ____ Street" or a similar form.

Do not describe the two portions of a split lot from opposite sides, as: "the east 50 feet," and "the west 50 feet"; use: "the east 50 feet," and "Lot____, except the east 50 feet," or a similar form. Otherwise there is every possibility that the parcels do not have a common boundary line, due to variant dimensions shown by relocation survey. On the record, all may be well, but in the fact, it is usually faulty. It is never safe to assume the record dimensions as correct except those which are given and interpreted as exact (such as a distance or bearing having no call or tie as an extremity).

Never use the directional form (1a, ante) of lot division (Ely 50 feet, etc.) which is intended to describe that portion of a corner lot having as a part boundary of the portion, the curved corner; from the definitions and interpretation, that part of the dividing line opposite the curve is indeterminate; (See notes on parallel lines and curves, and paragraphs following (d) above).

The following forms are acceptable for use in such cases:

Lot____ except the W 70 ft (which is the remainder).

Or; That portion of Lot____ lying Ely of a line parallel with and distant Wly 50 feet measured at right angles from the tangent portion of the East line of said lot.

Never use the directional form (1a, ante) when the boundary of the portion opposite the recited boundary of the lot is *not* intended to be parallel (see definition of parallel) with the recited boundary, in the aggregate of the parts of such boundary. (See Fig. 8)

Since there is *NO* true Ely line on a pointed lot whose Ely boundary consists of lines bearing greater than 45°, there can be no accurate location of "the E 50 ft"; the division line must be described with reference to some other line or part of the lot; this holds true also in cases where the Ely line of the lot is a curve whose ends are tangent to lines bearing more than 45°. (See Figs. 1 & 19). The rule holds true also if the radius of a circular segment of the Ely boundary is less than the perpendicular (right angle) distance between the lot boundary and that of the described portion.

Do NOT use the term "Fractional Lot" as is the frequent practice in assessment descriptions; it is vague and meaningless; the "fractional portional of a lot" exists, but a lot, as such, can not be fractional, except by definition as recited in sectional areas; if any part or several parts are cut from a lot, the remainder in either instance is considered "fractional" and uncertainty results.

The assumption that a lot line is changed by the conveyance (especially by a strip for street purposes, or a resubdivision of a part of the lot) of a portion, is incorrect; lot lines do not change from the original record showing; the title or usable portion of the lot may be reduced, but the lines are fixed, even though dimensional error prevents exact relocation. The showing on the map deposited in the Recorder's office and given a legal reference by book and page or document number, or both, on a certain date remains the official record and is not changed. 212

The most perfect form for division of a lot into two parts is the recital of the dividing line between the parts, described identically for each part, supplemented by the direction (E of or W of) of the part from the line; there is no possibility of conflict between such parts.

In describing portions of two lots, or all of one and a portion of another, note the following variants and interpretation:

portion of Lots A and B
  affects part of each lot

portion of Lot A and Lot B
  affects part of Lot A and *may* affect Lot B in part or cover all Lot B

portion of Lot A and *of* Lot B
  affects part of each lot

Lots A and B except portion—
  (or except portion thereof)
  uncertain; may affect either or each, or the aggregate (A & B)

Lot A and Lot B except portion—
  may affect A and B or B alone

Lot B except portion—and Lot A (or all Lot A)
  good and affects B only

All Lot A and Lot B except portion—
  construed usually as affecting B only as to the recited portion; should be used with discretion.

Lots A and B, except portion—of said Lot B
  best form and shows clear intent as affecting B only in the part.

(2) Directional division, more than 2 parcels from 1.

These forms are particularly useful in cases where it is desired to describe a number of parcels, having definite relation between themselves and whose sequence of use is unknown.

The use of successive distances, from the base line or corners, which are mathematically consistent, and the recital of parallelism with the same line, produce parcels which cannot be in conflict and which will leave no undescribed strips between.

To complete the division of the base area by the above methods, the remainder parcel, after all the previous ones have been taken out,

212 must be described as the whole, except the parts previously described.

An example of division of a 4 sided lot, 206 ft along the West line, 327 feet along the South line, having an angle of 90° at the SW corner, into 50 by 150 foot lots along the S line, a 170 foot depth lot at the NW corner and remainders would be described as follows:

1. The W 50 ft of the S 150 ft of Lot
2. The E 50 ft of the W 100 ft of the S 150 ft of Lot
3. The E 50 ft of the W 150 ft of the S 150 ft of Lot
4. The E 50 ft of the W 200 ft of the S 150 ft of Lot
5. The E 50 ft of the W 250 ft of the S 150 ft of Lot
6. The S 150 ft of Lot, except the W 250 ft thereof
7. The W 170 ft of Lot, except the S 150 ft thereof
8. Lot_____ except the S 150 ft and except the W 170 ft of remainder thereof.

Note that 6, 7, and 8 take up *all* remainders of the lot in the directions named, by making the exceptions of parts already considered, thereby describing, 1 to 8 inclusive, the *whole* lot without conflict and leaving no undescribed parts.

Care must be taken in the use of ". . . of the remainder of said Lot." (See Fig. 25)

If the exceptions, save the last, leave an irregular shaped parcel, the expression "except the Sly 80 *ft* of the remainder" (see diagram) may result in two interpretations. Use "also except the Sly 80 feet of said lot" or "also except the Sly 80 ft of that portion lying W of the E 105 ft thereof, of said lot."

The repeated exception of a parcel, or the exception of a parcel greater than necessary to affect the described land does no harm, as it cannot be misconstrued in its effect on the described land.

Frequently portions of a lot are conveyed, each part described from different parts of the boundary, such as

1. the E 60 ft; and
2. the W 40 ft of the S 100 ft.

A description of the portion between, then, must be described as "Lot_____ except the E 60 ft and except the W 40 ft"; or "the S 100 ft of Lot_____, except the E 60 ft and except the W 40 ft." etc.

Another occasionally used form of multiple division producing non-conflicting parcels is the following:

> That portion of Lot_____ lying between lines parallel with and distant Northerly 500 feet and 550 feet respectively measured at right angles from the South line of said lot and between lines parallel with and distant Westerly 300 feet and 450 feet respectively

from the E line of said lot, measured along the South line of said lot. 212

The fact that variance in dimensions of a relocated lot from the original record of it, due to physical changes, differences in Surveyors' measurements, realignment of streets, movement or destruction of monuments, etc., is of common occurrence, makes it absolutely necessary to maintain good non-conflicting title of parts of a larger parcel by using the exception form above set out for remainder portions.

Exception or partial inclusion form 213

Many instances arise wherein title is held in a certain parcel which is partially included within another, and it is desired to describe the overlapping areas. In the absence of survey, or of information regarding the dimensions of this area, the most satisfactory method of description is the partial inclusion type. It may be expressed in this or similar manner:

"That portion of the land described in deed recorded in Book ____ Page ____ of Deeds, included within the following boundaries (here describing the same)"; or

"That portion of the East 5 acres of Lot 6 of X Tract included within the West 200 feet of Lot A of Doe Subdivision"; or

"That portion of the land described in (or: conveyed by) deed to John Brown which does not lie within the boundaries of March Tract";

This latter form is frequently used to cover small strips of doubtful size or existence, along the boundaries of a larger parcel, and is effective for quitclaim purposes.

Although the priority of recital of either parcel over the other in the exclusion form is immaterial, it is preferable to give the first call to the property of the senior title, limiting its area by the second call.

The exception form of description is very common, especially when concerned with the exclusion of street areas.

It is also definitely of value and necessity in good title in part parcel descriptions (see above) wherein portions of the larger parcel have been conveyed and it is desired to deal with the remainder. It might be said that every remainder parcel description should be of the exception form, except those covered by metes and bounds.

The exception form is simple; first describe the whole area within which the instant parcel lies, then add the description of the parts to be left out, prefacing them by "EXCEPT . . . ," or "EXCEPT that portion thereof described as follows: . . . ". If there is more than one exception, prefer the sequence of priority, unless the intent is more clearly shown by some other order. When there is more than one exception, preface each one *after* the first one with the word "ALSO"; i.e., "ALSO EXCEPT . . . "

A double exception of the same area is not harmful, and almost never creates uncertainty; such as "Lot 6 except the East 30 feet and the South 50 feet"; in this case the four-sided figure 30 x 50 feet in the SE corner is excepted twice.

It is dangerous and very inadvisable to recite in a description an exception which itself has an exception. Almost invariably uncertainty results, and the description may be interpreted as covering *more* or *less*, than the area intended. In most cases the exception from the exception can be entirely omitted or carried only as an additional exception to the principal area. (Fig. 15) (Compare Fig. 25).

214 *Boundary recital form*

In this form the proper phrase following the caption is: ". . . bounded as follows," instead of the usual "described as follows"; then recite the boundaries of each side or sides, as, for example: "on the North by the land of John Doe described in deed recorded in Book 10 Page 66 of Official Records; on the East and southeast by Keen Street; on the South by the right of way of the M. P. & R. Railroad Co; and on the West by the various courses of the East line (or simply—the East line) of tract Q as per map recorded in Map Book 16 Page 199."

The order of recital of boundaries is immaterial although for convenience the most readily locatable boundary should come first, and the others successively clockwise or counter clockwise.

The frequently used old form of—"on the South by land of the grantor" is indeterminate as to such boundary unless it is otherwise specifically described.

The boundary form description eliminates almost entirely the need of dimensions, defining an area perfect in its relations with adjoining lands, without conflict, and dependent wholly upon the accuracy and locatability of the boundaries of such adjoiners. It is very suitable for areas not covered by survey, and for instruments of a preliminary or temporary holding character.

It is generally advisable to bound the extremities of strip descriptions such as are used in easements extending from the described property to a street or other easement, especially if the end lines of the easement are irregular or at an angle to the side lines. It has been held that, without qualification, the end of a strip is normal (at right angles or radially) to the end of the recited line.

See Section 216 (a) for further qualifications.

1. *Vicarious or quasi metes and bounds* 216

(a) Strip descriptions.

The strip description, which refers to the area by stating its width and giving its relation to the successive courses of a reference line, is a very common form for road or right of way conveyance.

This form is especially adapted to the "blanket" type of conveyance, in which the same whole description is used for the acquisition of any portion; any *overlapping* interests are wholly conveyed, as regards the strip, and the necessity of making accurate location of crossing property lines is largely eliminated; any subsequent location change of any portion of the strip other than that crossing the property of the grantor, may be made without a new deed from such grantor. The grantor may only convey the portion owned by him regardless of the excess area beyond such ownership which may be described; a statement to this effect is often made in the deed and is good, though not necessary. In the case of slivers between major parcels over which the strip passes, there is no automatic conveyance of the sliver; it must be covered by a separate deed.

If two strip descriptions from opposite directions meet in a common line, the tie of each to and in that line *must* be *identical* to prevent conflict; if the reference line of the strip ends at an angle point on a diagonally intersecting boundary, continue the description of the next course far enough to extend into the adjoining property to prevent gain or loss through extension or shortening of the side lines. (Figs. 9, 10, 11)

The side lines of the strip are presumed to be prolonged or shortened to terminate in the boundary of the land of the grantor, without recital, but such statements may be made in the deed if desired. An exception to this occurs when the reference line terminates *within* the area of the grantor, and no statement of extension is made; such a termination presumes the end line of the strip to be at right angles (or radially) to the course which terminates the strip. Another exception occurs when the boundary of the grantor's land contains an angle point of the strip; in this case the strip side lines are continued around the angles instead of being prolonged or shortened, and no statement of extension need be made.

The reference line is usually the center line of the strip, although any parallel line may be used.

The form of description may be as follows:

(1) A strip of land 50 feet in width, the center line being described as follows, etc.

216 (2) A strip of land 50 feet in width lying 20 feet northerly and 30 feet southerly from the following described line, etc.

(3) A strip of land 50 feet in width, the Northerly line of said strip being described as follows, etc.

(4) A strip of land 50 feet in width, the Easterly line thereof being parallel with and 30 feet Westerly from the following described line, etc.

(5) If the direction of the strip varies noticeably from a general straight line, such as one starting easterly, then turning southerly and southwesterly, then the above form (2) may be changed to read: a strip of land 50 ft. in width, lying 20 ft. to the left, and 30 ft. to the right, looking in the direction of the traverse, of the following described line.

It is correct to say "a strip of land 50 feet wide lying 25 feet on each side of the following described line" but it is incorrect to say "a strip of land 50 feet wide lying 25 feet on *either* side of the following described line"; "each" denotes the strip to lie on both sides of the line at all points; "either" implies that the strip may be on one side or the other.

Special forms may be used for strips of variable width or in which different portions are referred to different lines.

A description frequently used by utilities companies reciting "a right of way for pole line, the poles to be placed within one foot of the following described line . . . ," is to be considered as affecting a 2-foot strip, the center line being the described line; however, additional width may be gained by proper statement to that effect.

If the strips are augmented by rounded corners at cross lines or street intersections, these additions are best described as separate additional parcels. Such rounded corners may be satisfactorily described in form as follows: "a triangular parcel bounded as follows: on the east by the west line of N street; on the south by the north line of the above described 80-foot strip; and on the northwest by the arc of a circle of radius 30 feet, concave to the northwest, tangent to said west line of N Street and to the north line of said 80-foot strip."

It is not good practice to describe a parallel sided strip of any considerable length or number of courses by the metes and bounds method, because of difficulty in selecting some particular course for subsequent recital or tie, and because of bulk and cumbersomeness; nothing is gained by such metes and bounds, and an error in one line will disrupt practically the whole area.

It is always advisable to make specific recital of the end limits of the strip unless clearly evident. For example, a center line description beginning in the West line of a lot and extending easterly across several parcels and into another, terminating abruptly near the middle of the last area, need not carry the West boundary as "bounded by the West line of Lot ____" but should recite "bounded on the East by a line drawn at right angles (or a line bearing . . . , etc.) through the Easterly terminus of the last above described course." 216

(b) Description of boundary by recital, successively, of its segments by reference to matters of record, or dimensions, or both.

This form of metes and bounds is the most common form of general description used, and is applicable to the widest variety of use and circumstances.

A metes and bounds in successive courses may be perfect without a bearing or distance, provided every course or angle point is definitely recited as a matter of record, for example: Beginning at the SE corner of Lot 6 Tract M (place of record); thence NEly to the W corner of land of A (place of record); thence NEly along the NW line of said land of A to the S line of X Road 50 ft wide, etc.

The boundary recital form for part lot or parcel previously noted also can be classified under this heading.

2. *True Metes and Bounds* 217

In effect, any description may be accepted as a metes and bounds inasmuch as an area must have dimensions and boundaries for its determination; however, the term commonly applies to a description which recites successively the various courses of the boundary of the area, and the true metes and bounds must be so considered.

The true metes and bounds is the most perfect form of description possible for defining an area. It will include full caption, recital of all ties and monuments, whether record or physical, which determine the correct position of the boundary, all references to adjoining lands by name and record, and a full dimensional recital of the boundary courses in succession, which must be mathematically correct.

Such a description in its completeness is rare, and seldom necessary, but the amount of material incorporated in any case should be sufficient properly to identify the land described and make the title certain with respect to adjoining titles.

It is good practice, in a metes and bounds description containing a large number of courses, to successively number the courses for

217 ready reference; example: ". . . thence (1) N 27° E 460 ft; thence (2) S 88° E 1067 ft. . . . etc."

Comment on Metes and Bounds

In some jurisdictions (as in mining location patents), metes and bounds are required, with ties to physical monuments, with invalidity resulting if not followed, or at least, a refusal to record.

Some Governmental Authorities and financial organizations even require a metes and bounds description of a lot in a tract, conveyed as such, in addition to the lot number and tract reference.

The apparent advantage of metes and bounds, lies in the setting out of dimensions of the courses, and recital of monuments, giving the owner and others an idea of size, and an apparent fixity of position.

The rules of interpretation and precedence of calls, showing of discrepancy or error, physical destruction or movement of monuments, all combine to discredit much of the work involved in this form of description. As a general rule, an area may be more readily and specifically described by any of several forms other than true metes and bounds and with less liability of error or inconsistency.

Except in the case of free (no adjoiner) boundaries, the control rests with the record of previous titles and descriptions adjoining. The boundaries of the considered area must be held flexible to the extent of corresponding with previous adjoiners, and it is evident that a metes and bounds carrying dimensions together with boundary recitals is liable to misrepresentation if dimensions disagree with a proper relocation of boundary land.

A fully dimensioned description provides a great temptation to a surveyor or an owner to ignore the facts of title and adjoiners, thereby saving a little time, but often creating a lawsuit.

Physical monuments recited in a description should never be accepted as found on the ground, without full check as to originality and undisturbed position (whether from natural or artificial cause) or proper replacement. It cannot be presumed that monuments recited in a description, and presumably discovered in the field, will hold against other items of identification of boundary, unless position is definitely established; the court may decide to the contrary, however, if the circumstances so warrant.

Even though it requires more research and collateral data to use a description other than metes and bounds in location and to rely on a reference map for dimensions and monuments, the additional information given by such material will repay the locator far more in accuracy and reliability than the time and possible convenience achieved in the metes and bounds.

COORDINATES 218

Descriptions of land boundaries, for location and record, may include recital of coordinate values for any or all angle points of boundary or tie as supplemental.

The courts have determined the sufficiency and validity of descriptions which use coordinates as descriptive of or reference to points or corners; however, coordinate use in connection with other types of reference is regulated in some state laws and such regulation must be considered. California law, for example, relegates the legal value of coordinates to a secondary position holding that ties to legal boundaries are primary.

The purpose of such coordinate recital is the co-identification of physical and record points and precise relocatability thereof.

Every coordinate system must have a base. The usual base is a Plane Coordinate type, wherein monumented stations of the USC & GS triangulation network are given coordinates computed for projections of the stations onto a designated projection or development Plane (Lambert Conformal, Mercator, etc.).

From these projection coordinates the coordinates and position of any point within the base area may be obtained through physical ties and computation to the already established monuments.

The system in California is designated the California Coordinate System, Zone __ (there are 7), and description reference may be so made (see examples following).

Considerations

A description is the written record of a physical area or boundary, for the purpose of perpetuating location and title.

A description to be definitive must recite legal or physical monuments defining boundary, or associated therewith, and properly related to each other by dimensions which can be developed from survey or record or other reference.

A description, to be useful and fully operative must carry sufficient recitals other than, or in addition to, dimensions and statement of monuments, which will specifically determine the interest of the parties named in the document, and indicate the integrity of the land boundaries as related to claims of title or adjoiners.

It is a general rule that any description is good and valid (as such) which identifies and distinguishes a specific parcel of land from all others, regardless of title limitation or qualification, but such rule

218 may be modified or limited by legislative act for purposes of recordation and legal notice.

The lines and angles of boundary should be sufficiently described by dimension, reference ties, and record reference, to enable an accurate relocation in physical position, even if and when the original monuments are disturbed or destroyed, or if the original angle positions had no physical existence other than recitals in the description.

For transfer of title, based on record data, a valid and competent description can be written without dimension, the location and boundaries being fixed wholly by ties and references to points and lines already established in earlier records; or a combination of record references and supplemental dimensions may be used. If the parcel to be described contains or adjoins no record boundaries or corners, then the description must be wholly dimensional, but with at least one tie to a record reference point, for location and identification.

It is possible, and frequently occurs, that a parcel of land is described, and transferred by recorded deed, one or more times before any ground location is surveyed or monuments set. Under such circumstances there can be no recital or description of such monuments, and the given dimensions control over any subsequent monument positions unless there be a legal adjustment to the contrary.

Every description in which statement of distances and bearings are used, should recite the unit of measurement (feet, chains, etc.), and the basis of bearings (a line of record, true meridian through a specified point, grid meridian, etc.). The unit of measurement is usually given with each statement of quantity (75 feet, 23.44 chains). The basis of bearings may be stated, or implied by recital of the record bearing and call along a record line, or by use of record bearings shown on the subdivision within which the described property lies, or by stating an assumed bearing as base on a specific line (preferably the first or tie line) of the description. The specific statement of the basis of bearings is especially advisable if any of the bearings in the description are given as "North," "due North," "due East," etc., since there is a possibility of misconstruction and assumption that such calls are applicable to astronomical or geodetic meridian instead of planimetric development thereof.

Although use of azimuth, instead of bearing, is valid for description purposes, it is not recommended for general use; common and long-time practice of bearing recital has made such form more readily understood and acceptable, except in certain technical situations; since azimuth may be reckoned from either north or south, the need of statement of origin is evident in any description using it; azimuth generally is based on true meridian through the initial point of each

line recited, thus causing a variation between azimuth angles and 218
measured angles between lines; the recital that azimuths are given as angular relations from a single stated meridian, is necessary in any description involving the usual planimetric system.

The most accurate and satisfactory method of survey relocation of monument position is that of rectangular plane coordinates, developed from the triangulation control stations established by the U.S. Coast and Geodetic Survey, segregated for practical use into the various State Coordinate Systems.

Any monument position may be located and given its correct rectangular plane coordinate values in the applicable State Coordinate System. Such values, however, are inoperative, by themselves, for record relocation, unless defined or recited in an existing record. If a description or map of an original location, or of a properly determined relocation, defines the monument or its position, and gives also the coordinates thereof, such description or map is competent to serve as a base for subsequent location from the coordinates.

Inasmuch as the coordinate method is devised and used primarily as a means of accurate survey processes, its adaptation to title matters must necessarily be limited to a subordination in favor of other requirements of title and record location. In any combination of statements including recitals both of coordinate values and record calls, if a conflict appears, the record will prevail over the coordinate position, in interpretation; but the contrary may be shown.

The distance and azimuth, or bearing, between two coordinated points, as computed from the coordinates of such points, are correct only as grid dimensions; the actual length between the two points on the ground, and the bearing of the line between such points will differ from the grid dimensions; therefore in a description, any recital of coordinates for a point or monument must be specific as to its value and relationship with the true ground and title position and values of the point. The grid dimension will correspond to the ground dimension only when a balance or compensation exists between the elevation and scale factors used in transforming actual ground measurement to grid dimension.

Any point or ground position, monumented or not, can be placed by its coordinates, and replaced with precision by proper survey processes. It does not follow, however, that such a point is a correct title and record position. The surveyor who fixes the coordinates for a certain monument, claimed by him to be a corner of title, merely identifies an accurate procedure for relocation of that monument, but does not determine its title value. There is no criterion of the title accuracy of a point merely because it is given coordinate values.

218 Unless and until the coordinates of a point are established of record, either in description a priori, or by court order, as a superior call for a title fix position, the value and weight of coordinate recital is relegated to the same level as that of any other dimension.

Any person, proprietor, or court, interested in land boundary, is concerned with its title certainty and physical position; it is of no moment that the description recites coordinates or grid dimensions; these are merely more or less meaningless (to him) appendages accepted as routine or technical additions to the description toward making it a formal instrument. The surveyor making the ground location of the described coordinated corners, is the one who must recognize the value, possess the knowledge, and have the ability to apply coordinate dimensions. If he has the "know how" to add Coordinate System technique to ordinary surveying procedure, a proper location is possible; otherwise he had better leave it alone. Under such circumstances the client is advised to rely on legal interpretation of his title and possession.

A valid and good description may be written by reference, partly or wholly, to a map of record. Proper reference to such a map reads into the description all matters shown on the map which may be pertinent to the location and interpretation of the boundaries of the described parcel; such matters may apply to the relative position and size of the parcel indicated by the plan of the entire subdivision, or even extend to the exterior boundary dimensions and ties thereof.

Maps are a valuable asset in description reference because of their capacity for indicating multiple ties, delineation of associated land, extra dimension and location data, coordinate values of points and monuments, and notes of general character, any or all of which may have a tendency to involve and confuse the worded description. For title and reference purposes it is essential that the map be recorded in the proper public office, since only such office can lay good claim to the giving of legal notice to the world at large.

Examples

The following examples and comments will illustrate the use and application of coordinates applied to descriptions usable for title purposes.

Example 1: A parcel of land in Black County, State of Q, described as follows: Beginning at a 4 x 4 stake in a mound of rock in the north side of Highway 2, said stake having grid coordinates X___, Y___, of Zone 5 of the Q State Coordinate System; thence northerly to a point in Ash Creek having grid coordinates X___, Y___, of said Zone 5; thence easterly to a 2″ pipe marked "MH," having grid coordinates X__, Y__, of said Zone 5; thence southerly to a point having grid coordinates X__, Y__, of said Zone 5; thence westerly to the point of beginning. 218

COMMENT 1

If we accept the definition of specific locatability and identification of position, this is a perfect description, since there is only one possible location of the described land.

A legal restriction, permitting the use of coordinates only as an addition to record references, would tend to make this description legally invalid or voidable.

If the described area has any record title boundaries, the description here fails by non recital of such boundaries.

The recital of the State Coordinate System (and Zone number, if any) as here noted, is sufficient to identify the matters related thereto, including the coordinate values and the mathematical processes necessary to fix the monument positions; also both the grid and actual dimensions. However, technical knowledge is necessary to make such determination, making the description as it stands, of little value for purposes of assessment, taxes, or even ground boundary dimensions and associated properties.

EXAMPLE 2: A parcel of land in Black County, State of Q, described as follows: Beginning at the southeast corner of the land of Doe as described in deed recorded in Book 27 Page 122 of Deeds in the Office of the Recorder of said County, said corner being marked by a 4 x 4 stake in a mound of rock in the north line of Highway 2, said stake having established grid coordinates as recited in said deed to Doe, of X__, Y__, of Zone 5 of Q State Coordinate System; thence along the east line of said land of Doe, N 15° 37′ 22″ W, (bearings based on the grid meridian of said Zone 5), a distance of 473.88 feet to the northeast corner of said land of Doe, a point in Ash Creek, having grid coordinates X__, Y__, of said Zone 5; thence N 65° 30′ E 698.70 feet to a 2″ pipe marked "MH" set for the southwest corner of Lot 4 of Tract RR, as per map recorded in Map Book 10 Page 45, Records of said County, the grid coordinates of said pipe being determined in this description as X__, Y__, of said Zone 5; thence S 10° 28′ 55″ E to the north line of said Highway 2; thence westerly along said Highway to the point of beginning.

COMMENT 2

218 This description is good and sufficient to fill title and location requirements when full use is made of the recitals and ties given.

The coordinate values allow of a correct repositioning of the designated corners, if destroyed or lost or unmonumented.

The base of bearings is definitely stated, allowing comparison with bearings from some other base, if both are referred to a common line of description; also allowing check of measured angles against bearing differences.

The coordinates are recited in a manner making them subordinate to title calls or dimensions, if a difference exists.

The location of Highway 2 has to be determined from the deed or map or field notes thereof which are matters of record; the description here permits a flexible connection but a rigid contact.

The given bearing from the coordinated 2″ pipe to the Highway provides a fixed direction correctly related angularly to course one of the description.

EXAMPLE 3: That portion of Lot 6 Tract MM in the County of Black, State of Q, as per map recorded as filed map No. 435 in the office of the Recorder of said County, described as follows: Beginning at the most easterly corner of said lot; thence northwesterly along the northeast line of said lot a distance of 378.55 feet; thence westerly in a direct line 843 feet more or less to a 1″ iron pipe set to mark the northeast corner of land of Roe as described in deed recorded in Book 19 Page 88 of Deeds, Records of said County, said pipe having grid coordinates X__, Y__, of Zone 5 of Q State Coordinate System, said coordinates being determined for said pipe in this description; a 3 x 3 stake intended to mark the northeast corner of said land of Roe, but discovered to have been removed and erroneously reset lies northerly 1.75 feet, and has grid coordinates X__, Y__, of said Zone 5; thence southerly along the east line of the land of Roe to the southeast line of said Lot 6; thence northeasterly along said southeast line to the point of beginning.

COMMENT 3

This description is complete from a title standpoint, but requires record search by the surveyor to dimension the boundaries.

Examination of the description shows the intent to be principally a fixing of the northerly line of the land in question; the east line is a lot line determinable from the map data, with the east terminus of the north line of property fixed at 378.55 feet along the line from the easterly corner of the lot; the west terminus of the north line is fixed in the description by monument and coordinates, which here are made paramount in event of destruction of the pipe; the intent seems to be to fix the pipe as Roe's true corner as against the disputed location of the 3 x 3; the west line is the east line of Roe, a good title tie; the south line is a boundary of the lot. 218

It is evident that it was considered unnecessary to coordinate any points except as noted; quite possibly the map of Tract MM showed coordinates for its corners, or some other specific means of their location was available.

EXAMPLE 4: The Southeast Quarter of Section 17 T 3 S, R 5 W,–Principal Meridian, in the County of Black, State of Q, the corners of said Quarter being more particularly defined as follows: The SE corner an original stone in mound, having grid coordinates X___, Y___, of Zone 2 of Q State Coordinate System; the East Quarter corner of Section located on line midway between original SE corner and original NE corner of said Section, and having grid coordinates X___, Y___, of said Zone 2; the South Quarter corner of said Section, a 2″ capped pipe set for said corner by the Surveyor of said County, having grid coordinates X___, Y___, of said Zone 2; the center of said Section, a 4 x 4 stake in mound of rock, set at intersection of straight lines connecting opposite quarter corners of said Section, having grid coordinates X___, Y___, of said Zone 2.

COMMENT 4

The first sentence of the description is sufficient for any title purpose, and needs no qualification. The addition of described corners and coordinates furnishes a definite means of relocation in the event any or all the monuments are altered or destroyed. The recited coordinates will also furnish information as to any inaccuracy of described location for the east Quarter corner and the center of Section, if the lines of Section are rerun and show discrepancy. Here as in other examples, the coordinates are confirmatory only, of the correct corner positions, the description of the corner taking precedence of the coordinates as long as the corner may exist.

The description above could also be written as a regular metes and bounds, reciting point of beginning, traversing around boundaries, reciting monuments and coordinates, etc., but gaining little, if any, advantage in space or clarity.

218 Example 5: That portion of Block 14 of the Townsite of KK in the City of Green, County of Black, State of Q, as per Map No. 999 Records of said County, described as follows: Beginning at the intersection of the center lines of Main Street and 10th Street, in said City, a point having grid coordinates X__, Y__, Zone 4 of Q State Coordinate System, said intersection and center lines and coordinates being established by the City Engineer of said City and shown on Street Map No. 6354 on file in the office of said Engineer; thence westerly along the center line of Main Street 127.99 feet; thence northeasterly in a direct line to a point in the center line of 10th Street distant northerly 87.00 feet from the point of beginning; thence southerly along the center line of 10th Street to the point of beginning; except those portions within the lines of said streets.

Comment 5

Reference to the City Engineer Map furnishes all the additional information needed, outside the description. It is very effective practice for the City Engineer to establish street center lines, coordinate the intersections on the State System, make a map showing all data, and file it in his office; recordation is unnecessary. The exception of part in streets effects a description of the part of block only, and saves extra calls and dimensions required to bound the part in the block; the surveyor must make the computations in any event, if he is to stake the land.

Example 6: That portion of __ (here follow with the ordinary type of description, part lot, or metes and bounds, etc. and then add __) For more particular description and location of the boundaries of the herein described land, reference is made to the map attached to this instrument and made a part thereof.

Comment 6

The vital part of this example is the reference to map attached which map must be recorded as a part of the document. On this map can and should be shown the platted boundaries, description of found and set monuments, coordinates of at least two corners, dimensions between all corners and ties, note as to source or base of coordinates, notes, or references to any applicable record data of deeds, maps, etc.

An alternate form of the map attached is that of recording, separately and priorly to the instrument containing the description to be identified, a map of the area in question, following the legal procedure required in the district for such recording, and showing on such map all necessary information as above outlined; then the description carries the reference to that map recorded in— etc., instead of "map attached." 218

EXAMPLE 7: — — — to a 2″ pipe with tag (L.S.666), having grid coordinates X__, Y__, said Zone 3; thence along the east line of said Tract WW, N 26° 47′ E (base being grid meridian of said Zone 3) (record bearing of said east line Tract WW being N 26° 33′ E) a distance of 487.00 feet to a 6 x 6 squared stone flush with surface, having coordinates X__, Y__, said Zone 3; thence S 76° 50′ E 333.00 feet; thence — — —

COMMENT 7

It is often advisable to show relation between different bearings on the same line for title or record comparison. There is no simple phrasing adequate to express the relation, so each case will present its own solution.

The bearing S 76° 50′ E follows the grid base with the proper angle to N 26° 47′ E; the record bearing for east line tract is informational only.

Conclusions

Descriptions written for title purposes may, in addition to the usual ties, recitals and references, give the coordinates of one or more corners or monuments of the described property; the coordinate values are supplemental to the calls in the description and provide a means of accurate physical location of the points determined by title recitals; coordinate values normally are subordinate to title calls, but may under some circumstances, take precedence over other recitals.

Reference to coordinates must be definite, specifying the type (grid), the Zone (if there be such), and the Coordinate System of reference; the Coordinate System, if established by legislative act, is sufficiently identified by naming the System as such; if not a legalized system, then more particular statement or reference must be made for identification.

"Grid" coordinates are preferred, as they are uniform in derivation in any particular system, and from them the surveyor may readily develop actual ground dimensions for any elevation or position in the Zone.

Base of bearings from an adjoining record bearing is usually most

218 convenient and acceptable in title, but any other is usable if sufficiently identified; grid meridian base, if used, should be compared or related in the description with the record bearing on some described line.

It is better where possible, to incorporate recital of grid coordinates in the body of the description, rather than place it at the beginning or end; the reason being that careless or intentional deletion of recital is more difficult and apparent.

If the description is complete as to dimensions, and is mathematically correct in closure, one coordinated point is sufficient to position property, but two points at some distance apart are better for check; it is rare that all points need their coordinates given.

If the description is incomplete as to dimensions, those missing covered by title references, then it may be advisable to give coordinates for all undimensioned positions.

No coordinates can be given for a point (A) unless there has been an actual connecting survey therefrom to a known coordinated point (B), or unless it is possible accurately to compute position of point (A) from a record coordinated point (C).

Because of the precision requirement of coordinates, none should be stated in a description until verified by double original check of the surveyor and an independent person such as County Surveyor or City Engineer.

The examples given above are not intended to set a rigid form of description but to indicate the variables and possibles of incorporating reference of coordinates into description; any form is acceptable which will show the clear relation between the items to be considered, the intent of the parties, the maintenance of title, the mathematical accuracy of dimension between points and positions, and those other matters necessary for proper interpretation.

NOTE: The subject matter of this Section with some modification to conform with the format of the syllabus, is reproduced from a paper on "Coordinates in Title Description" published in the January-March 1952 issue of "Surveying and Mapping," the official Journal of the American Congress on Surveying and Mapping."

W. C. W.

In general these forms will be quasi metes and bounds, or the directional division type, frequently divided into more than one parcel.

There is no particular form set up for these descriptions, the style being developed from the matters to be covered.

The essential features here are coordination of the parts, definite statement of the intent of each, certainty of boundary as far as the record will allow, recitals or dimensions which will prevent overlaps or strips between the parts or boundaries, simplification in phrasing, and locatability in the field; the exception to locatability will lie in those descriptions which in part must follow an uncertain record.

Among irregular and special form descriptions there are the following general types:

(a) Courses of boundary related by angles, deflection angles, irregular form bearings, magnetic bearings, mariner's compass bearings.

Liable to uncertainty of direction unless the form and/or direction of the angular measurement is clearly stated.

(b) Dimensions of differing kind, odd, or unusual units of measure.

Sometimes necessary to conform with record to use different units in the same description such as feet, chains, varas, rods, poles, acres, hectares, etc. The unit used should be stated for each dimension (e.g., . . . thence East 50 feet; thence South 6 chains, . . . containing 70500 sq. ft., etc.).

(c) General direction recitals, ties or measurements, reference to impermanent physical objects; indeterminate forms creating uncertainty.

Dangerous to use, and decidedly not good practice, even for preliminary work. Practical only as information to the person who is to write a correct description for documents to cover the area approximately stated.

(d) Approximations or reference to a general plan providing for specific positioning and description at some future time.

Subject to the same criticism as (c).

(e) House numbers and street locations.

A common method used by persons whose principal information is the address location and who consider the address as all inclusive of the physically existent property, both as to the title and record.

219 Very dangerous to use without supplementation, although the courts have held such descriptions valid in certain cases.

The possibility of arbitrary change of the house number or the street name, or both, and subsequent use of the old number and/or name in a document, without proper reference as to dates, makes the description uncertain; likewise, the presumed physical location so designated, carries no assurance of record title position.

Not recommended as a description form for record conveyance.

A variant of the number and street type, also unsatisfactory (the record must be searched for identification) is the form ". . . the premises (or house and lot) of John Doe, in X City, State. . . ." John Doe may own more than one place.

(f) Distances only from designated fixed points to determine portions of boundary in question.

For Example: . . . thence Nly 150 ft to a point distant Wly 60 ft from the NW corner of Lot 25 . . . ; or . . . Nly along a line passing through a point distant S 80° W 40 ft from the SW corner of Lot 10, and through a point distant N 67° E 175 ft from the most Ely corner of the land of A (deed reference), to an intersection with the N line of Q Street. . . .

Certain combinations of direction may produce uncertainty or double interpretation; watch this form carefully when using it.

(g) Multiple parcels using various forms of description for segregation of vested interest including common use easements, etc.

Many descriptions are best divided into parcels for clarity; or the separate parcels may have been acquired from various sources and do not fit one another as to boundaries, rendering an aggregate into one parcel inadvisable. If definite inclusion of all or part of one parcel is made in the description of another, it is advisable to make an exception phrase of such included part, to prevent the effect of a double description of the area; this does not apply to double exception form previously discussed.

It is often advisable for the grantee to take title not only to the parcel of land which can be readily and specifically described, but also to any vested interest of the grantor in other lands, particularly those adjoining the described parcel which may or may not automatically pass with the fee of the instant parcel; such vested interests may be easement rights over other privately owned land, attached or reversionary rights in adjoining streets or rights of way, or the interest in adjoining streets over which the public easement still exists, or has been vacated.

The inclusion of these interests in the main description may be difficult or impossible. The advisable method is to set them out in a separate paragraph or parcel, somewhat in the following form: **219**

> "together with (or, Parcel 2;) all right, title, and interest in and to M Street adjoining the above described land on the west, and to the west 15 feet of the alley, vacated, adjoining said land on the east, and any reversionary rights accruing to said land by reason of the general plan of restrictions affecting said Tract X, as recited in deed recorded in Book ____ page ____ of Official Records (or Deeds) in the office of the Recorder of said County, and those certain general and specific easements appurtenant to said land as granted by various deeds of record."

The call for the extra tenements as above set out is very advisable in cases of condemnation of fee for freeways or other lands for public use, where some portion of the area is occupied by public roads or ways, not vacated, and portions of the underlying fee thereof are uncertain as to nature and extent of vesting. Of course, it is necessary to make defendant *all* persons having any possible interest in these lands or interests.

Exceptions or grants of interests of various sorts are also better segregated and set out from the main part of the description.

In acquisition of freeway areas, the rights of access to the freeway are taken from the grantors; the deeds carry the grant as a separate item, such as: "... also all rights of access to or from said parcel above described ...."

If the granted area includes land deeded for street purposes, which deed is questionable as to its status as fee or easement, and all interest of grantor is to be conveyed, the safe form of exception is: "... Except the interest of the (County of Los Angeles) in that portion of said land in M Street as conveyed by deed recorded in (deed reference) ...."

If a partial interest in a utility and the necessary easements and land is to be withheld by the grantor, a form similar to the following may be used: "... Excepting (or reserving) a two thirds interest in and to the well and pumping plant, and the pipe line distributing system located on said land; also excepting the land occupied by said well and plant described as follows (description of parcel); also reserving easements for such distributing system over strips of land ____ feet wide, the center line of said easements being described as follows (describe the easements)."

(h) Multiple parcels using only one caption.

The danger of this is that one of the parcels may be "lifted out" and used separately without a caption. In such a case the information could be insufficient for legal conveyance of title.

219 (i) Tax and Assessment descriptions.

These are generally drawn from maps prepared almost wholly from records, with or without certain adjustments to later surveys; such maps are rarely based on direct survey, except as an adjunct to the record. *Very dangerous* to use for relocation or conveyance purposes because of conflict in records, and the fact that the descriptions are usually deficient in recitals or ties to proper and complete record and rarely follow the precise language of the description in the record, thereby causing a variance in interpretation of boundary location.

Descriptions should *not* be drawn from such maps without additional qualification of the record; any descriptions made for assessment purposes must be used for informational purposes only, or applied as indicating the intent of the Assessment Agency to cover parcels which have been specifically described in the records.

Maps of the Assessor's records should *never* be used as reference in descriptions, except as an explanatory or secondary recital, and then, *only* if the description is complete without such reference; however, it is competent to refer directly to "Parcel No. ____ as shown on Assessment Diagram filed in connection with the opening and widening (or condemnation) of ____ Street in Case No. ____ Superior Court of said County."

The term "Fractional Lot" as used in such diagrams and descriptions, is incorrect. There can be *no* "fractional lot"; a "fractional portion" of a lot is possible, but a *whole* lot can never be *fractional.* The term is erroneously used indiscriminately, whether only one or several parts of the lot have been severed from the assessed area, by any or several methods; this makes the term indefinite unless examined particularly in connection with the special map from which the assessment description is drawn.

(j) Prism and Elevation descriptions.

When properties, such as tunnel volumes, or bridge areas, or floor areas of buildings, or stairway volumes, or airway easement areas and volumes, are described, a third dimension, elevation, is required.

In prism descriptions (1) the plane area, or projection of it to ground level is always necessary and the boundaries of such area are handled in the same manner as in any described area; (2) the elevations of the control points are given together with the horizontal dimensions necessary to fix their position with reference to the plane area; (3) statements as to the nature of the grant, its uses, and other information as to the relation of the grant to the volume described are incorporated in the description if convenient, or in the body of the instrument.

Elevations are given in feet above or below a stated datum; if the datum be established by Ordinance, reference is so made; if datum be assumed, or if elevations are made as direct differences to a physical monument (corner of walk, top of curb, etc.) the facts are stated. 219

In a case such as a conveyance of easement over a bridge, but not over the ground beneath, the description should cover the plane area of the exterior dimensions of the bridge structure, "above an elevation of ______ feet, ______ datum," the elevation being that of the *lowest* point of the bridge structure (except piers), as one parcel, together with descriptions in the regular manner (no elevations) for the areas covered by the piers and foundations, as other parcels.

Prism descriptions have no standard form, but must be constructed to suit the case in question, to insure certainty of interpretation.

(k) Descriptions including street or other areas subject to easement or vacation conditions.
(See section on easements)

(l) Descriptions of areas adjoining tide lands, shores, and water courses.
(See tide land section)

(m) Mining location description.
(See mining location)

Every description must refer to one or more monuments, whether legal or physical. The reference to physical monuments is often omitted and may not adversely affect the interpretation of the description, particularly if the record references carry recitals of physical monuments, or the description implies the existence of them.

Properly, any monument recited in a description should carry the notation of "found" or "set"; the description of the kind, condition, and markings; the status, whether "the corner of," "set to mark," "accepted as," or other statement showing relation of the monument to the described boundaries; unless the facts are clear from the context, or the reference map.

There is no set form of recital, but the following are the more common:

1. —to found original (*desc. mon.*);

2. —to a (*desc. mon.*) set as a replacement of the original, disturbed (or destroyed);

3. —to a found (*desc. mon.*), being a replacement of the original monument;

4. —to a found (*desc. mon.*) accepted as marking the corner—;

5. —to a (*desc. mon.*) set for the corner— (or, —set for the position of the corner—);

6. —to a (*desc. mon.*) set to mark (or, designate) the corner—;

7. —(*direction*)—(*distance*)— to a point at which set (*desc. mon.*); thence—;

8. —to a (*desc. mon.*), set (*by*—, if known), for the corner of—;

9. —to a (*desc. mon.*), set to mark the corner—;

Comment

Note the variance in interpretation placeable on the different forms above; use the form applicable to the particular problem.

In #7 the surveyor "sets," not "finds," the monument.

#8 implies the acceptance of the corner "set by ______," as the correct corner.

#9 implies the monument "set for" the corner by the previous surveyor, possibly only marks or witnesses the true corner position, if other items of the location indicate the contrary; at least the acceptance is questionable. Compare #5 above.

Another form, and quite common practice, is this: –thence––– 220
to a 3x3 stake; thence ––– to a 1″ pipe; thence–etc.

In describing a free boundary (page 21) this form is good because it makes the monument hold the position.

However, concerning an already established title boundary, the procedure of full description should also be followed when a monument is recited as a reference, whether it be on the instant boundary, or offset, or used as a specific tie point.

300 SECTIONAL PROPERTY

301 *Government Survey Sections*

The division of the public lands into sections and other areas as well as all locations of section and quarter section corners, or of the subdivisions of sections, are governed by application of the rules and procedure laid down in the Manual of Instructions for the Survey of Public Lands, issued by the Bureau of Land Management, Department of the Interior (formerly General Land Office) of the United States. This applies to original corners or to title lines of patents from the United States; modification may be made in some instances after the land has passed into private ownership. State laws and court decisions governing relocation of corners of patented lands may depart from the Manual provisions.

Original corners, or determined or identified replacements thereof are to be accepted as such regardless of differences in dimension to or from such corners. Other corners are located by certain fixed rules as laid down in the manual for establishment of interior and subdivision corners, and for lost and obliterated corners. Resurveys and new locations of section lines within or on boundary of Government-owned lands, although authorized by the Bureau, are not to be accepted in preference to the original unless they are definitely approved by the Bureau in lieu of the original (similarly to the practice of a subdivision owner who may replat a record subdivision in another form and record the new plat), or in certain special instances; such new locations are only applicable to public unpatented lands or to adjoining patented areas in which a proper legal adjustment has been made with the patentees.

For corner and monument location see Sec. 704.

After sectional land has passed into private ownership by patent, the method of subsequent subdivision of said land becomes a matter of intent and option, the same as any other privately owned land; it is only custom, good practice and assumption of probable intent, that allows locations of lesser parcels to be made according to the rules of subdivision as provided in the manual, unless the description uses a form of recital such as: "SE ¼ of the W ½ of the NE ¼ of Sec. ____ according to the Government Survey (or Township Plat) thereof," in which case the reference to "Government Survey" definitely determines the method.

Proration for a corner near a Range or Township line, using the nearest corners, one of which is across such line in the adjoining

Township, is incorrect and unacceptable; the Range or Township line must be retraced, and the nearest corner to that to be relocated is established; then the desired corner is prorated from that Range or Township line corner, to the next on the opposite side.

If the Field Notes show a free line in a Township with no connection with the next successive corner (Example: –from the Corner common to Sections 23, 24, 25, 26; east 40 chains; set–/desc. cor./–), or any extension and rerun from the closing (already set) corner, the proper method of relocation of such free corner is to run the record as set out in the Notes, i.e., east 40 chains. If the line between the adjoining corners is substantially east, permitting only a small side displacement of the intermediate corner, it will probably be advisable to make the three corners align; however, it is inadvisable to prorate the intermediate corner for distance, the record as given being safer.

It is unacceptable to prorate across unsurveyed portions of a Township (such as a survey covering the west 4 miles, and the east one mile of a Township, with no survey covering the mile strip between); the lines of the surveyed portions along the unsurveyed strip should be treated, for prorate purposes, as exterior lines of the Township.

A corner set by a Government surveyor within a patented (non Government owned) portion of a Township, as an incidental to other corners established by him in the Township, under instructions, and which is consistent with such other corners, is preferably used in relocation of such other corners, when the same are lost, providing the relocation is made by a private surveyor for privately owned section land, even though the Government has not approved the location or setting of such corner (since it is not on public land).

The center quarter corner of a section is fixed by the manual as the intersection of straight lines joining the opposite quarter section corners on the boundary of the section and has only *one* location. Any other corner set for the center may develop into a *title* corner but cannot correctly be said to be the center quarter corner. This condition may be altered as defined in the next paragraph.

There are instances in the original survey of sections, where the government surveyor has established other than the section and exterior quarter section corners, such as quarter-quarter or center section monuments, in surveys of fractions of a section; the approved Field Notes will disclose such corners and the monuments so set will determine and govern the positions of these corners regardless of the rules of location for secondary corners.

The location, and monumentation, of lines and corners by the

301 Government in a "dependent" resurvey, may or may not be acceptable as defining the boundaries of previously patented sections or ranch lands. The question of acceptability hinges on the accuracy of relocation of the original lines affected, the sufficiency of the methods used in such relocation, the possible conflict with possessory boundaries created in good faith by the patentees, and any adjustment of boundary between the public and the patentee subsequent to the Government "dependent" resurvey. Full investigation and survey by the private surveyor are necessary to show the sufficiency of the relocated Government corners before accepting such corners for any boundary positions in the area of survey. Descriptions in such an area should be very specific and complete in monument reference and ties.

The areas shown on Township Plats are computed from field note dimensions and are not determinative of boundary. Their use is primarily for indicating relative value or extent of grant. In patent descriptions, the acreage is subordinated to the other description, and usually quoted as "more or less" or "according to plat. . . ."

It is not correct to assume "E ½ of NE ¼" to be synonymous with "E 80 acres of NE ¼" (or kindred similarities) unless the identity is expressed in the description by reference or some other fashion; this is especially true if the NE ¼ was patented as a whole.

Likewise it is not correct to assume a numbered lot in a section as being identical with the corresponding quarter-quarter unless such identity is specifically expressed.

Questions of overlaps and strips on sectional land must be handled as in other property, except that the fractional area descriptions (E ½ NE ¼; SW ¼ SE ¼ etc.) are located by the rules of the Manual, instead of by actual areas; the nomenclature is, in effect, similar to a parcel designation rather than actual area.

Manual rules must be observed in fixing corners of irregular lots and subdivisions of sections on North and West tiers of Townships, and in Fractional Sections. Special cases, where the Manual is silent, must be solved by majority probability, endeavoring to arrive at a solution most nearly conformable with the intent of Manual operation.

302 *Fractional Government Sections*

The term "Fractional Section" defines any section of a Government survey which is not of regulation size, due to exception of portions which are otherwise held or designated, such as previous ranch or other grants, bodies of water, etc. It may be applied to sections which carry excess areas or lots over the regular division. 302

The division of fractional sections into their various subdivisions is subject to several types of analysis in determining those portions adjoining the areas excepted from the section, due to the missing (not set) exterior corners of the (whole) section.

For example, an E-W quarter section line, one end known, may be extended: (a) East (straight line at right angles to meridian through the ¼ corner); (b) easterly parallel with the N (or S) section line (between the section corners) or with the segment between the section corner and the quarter corner on the N (or S) line; (c) on an average bearing between the bearings of the N and S Section line or accepted segments; (d) by completing the section boundaries in the regulation manner from the existing monuments and extending the quarter section line from the existing corner Easterly to the theoretically determined opposite quarter corner.

The method employed will depend on conditions encountered and on the closest interpretation to the Manual.

Although the "S.E. corner" of the section may be evident as the intersection of the South line of section with the ranch line, it is good practice so to state, such as . . . "the S.E. corner of said section in the NW line of Rancho X"; it is especially necessary if there is an area adjoining in the ranch not conformable in boundary with the Section but designated by the same section number and being an approximate completion of the Government Section.

In the event the ranch line cuts off a triangular part of the S.E. corner of a regular section, the reference to "S.E. corner of said section" as above noted should be changed to read "Most southerly (or most northerly) S.E. corner–" according to location of designated corner being on the south (or east) line of the section.

Sections which are not of standard dimension usually carry the irregular portions as lots. The Manual prescribes the system of layout and numbering, but this is often altered to fit special forms and problems; the only certain determination of the lot number of an area is reference to and examination of the approved Township Survey Plat.

There are supplemental plats of portions of sections authorized by the Land Office; these bear lots numbered successively with those already existent in the section replatted. The dimensions are copied from the original section plat and the areas sum up to the total area of the original. No monuments are set and relocation is made by proportional methods only. Lots are not referred to as "fractional."

Original township surveys show numbered areas within the exterior boundaries of the township covering those lands to be excluded and not subdivided; these numbers begin with "Lot 37" and

302 are numbered successively; the areas are known as Lots in the Township. A Township Lot cutting a Section reduces the Section to a fractional one and the corners of the Section against such Lot are on the boundaries of the Lot.

Closing corner monuments on fractional section lines are determinative for line of the section, but *not* for line of the intersected grant; the grant line must be fixed independently and the section lines run to it on the line of the closing corner. However, for prorating distances along the section line, the record and measured distances for prorate are taken to the closing corner monuments; the distance between the actual fractional section line and the nearest prorated corner (along the intersecting section line) will be the prorate length corrected by the distance between the fractional section line and the closing corner monument.

Forest Reserve (Federal) corners are set by the Department of Forestry on lines and angle points of the Forest Reserve boundaries; where section or grant corners are found, the Reserve corners may be set 6 or 10 links from the found monument in the direction of the following course. Sometimes intermediate line monuments are set between angle points; they are not acceptable for determination of true line.

Field notes of the survey should be available in the district office.

Any Forest Reserve monuments found shall be tied in, but except for the information of position, they should never be used in land location on account of uncertainty.

The smallest legal subdivision recited in the Manual is the quarter-quarter section, and its corners are designated as quarter-quarter corners or sixteenth corners. However, the implied use and terminology of smaller subdivision areas result in the common forms such as "E ½ of the SE ¼ of the NW ¼ of the SW ¼," and the "NE corner of the SE ¼ of the E ½ of the SW ¼" etc.

303 *Descriptions in Sections*

Descriptions of sections as a whole follow the form of the Manual, such as "Sec. 6 Twp. 8 N, Range 16 W, S.B.M" or "Fractional Sec. 35 Twp 1 N, Range 11 W SBM."

"SBM" (or "MDM" etc.) in Section descriptions, to designate the meridian of reference, is correct; "SB*B*&M," while not wholly incorrect, is not in accord with the nomenclature of the Manual.

Descriptions of parts of sections first recite the portion taken, then the next larger portion out of which the first is cut, then successively the next larger portions of which the previous is a part, until the whole section is stated, with its Twp., Range and Meridian, as for example: "the E ½ of the SE ¼ of the SW ¼ of the W ½ of the NW ¼ of Sec. 10, Twp. 7 N, Range 14 W, S.B.M."

In a description reciting several portions of a Section or of part of a Section, it is bad practice, and a hazard in interpretation to condense the description of any portion and attempt to cover all portions by only one recital of the major quarter section area, such as in the following example. (Fig. 23)

(1) As intended:

(a) The N½ of the SE¼ of the NE¼

(b) The SW¼ of the NE¼

(c) The S½ of the NW¼ of the NE¼

all in Section 15

(2) Also correct:

The N½ of the SE¼ of the NE¼ and the SW¼ of the NE¼ and the S½ of the NW¼ of the NE¼ of Sec. 15.

(3) Also correct:

Those portions of the NE¼ of Sec. 15 described as the N½ of the SE¼ and the SW¼ and the S½ of the NW¼.

(4) *Incorrect*:

The N½ of the SE¼, the SW¼, and the S½ of the NW¼ of the NE¼ of Sec. 15.

This is very hazardous and permits of double interpretation.

Numbered lots in sections may be designated and described either by numbers, or by the quarter-quarter section or half-quarter section; the number is preferable and in most instances necessary; where uncertainty may exist, specify both the number and the portion of the section in which located, as—"Lot 4 in the NW ¼ (or, Lot 4, being the NW ¼ of the NW ¼) of Sec. ____" etc. Since lot numbers are given to those half quarters or quarter quarters of Sections which are irregular in dimension or area, the analogous call (Lot 4; NW ¼ of NW ¼) of the quarter quarter is doubtful from a title analysis, especially on subsequent division of the parcel (E ½ Lot 4; E ½ NW ¼ NW ¼) under private ownership. The safe way is to specify the method of division ("according to Township Plat" etc.).

In those cases where the Government Plat shows lots of the same number covering areas in different quarters of the same location, it is necessary to recite the part taken, such as: "Lot 2 in the SW quarter

303 of Section –," or "the W one-half of Lot 1 in the NE quarter of Section –," etc.

It is not correct to assume that "the E ½ of the SE ¼," "the E 80 ac of the SE ¼," and "the E 20 chains of the SE ¼," cover the same land, even though the section shows regular dimensions and areas of record, for actual resurveys of sections almost invariably show a discrepancy from the record. However, a statement such as "the E 80 ac of the SE ¼, shown as per Government Survey as the E ½ of the SE ¼," or some similar form of identification will interpret the description as covering the E ½ of the SE ¼ divided in accord with the Manual rule.

304 *Section Division in Ranchos*

Many of the Rancho Grants have been subdivided into areas dimensioned, named and numbered similarly to Government Sections. These areas must be considered in the same category as lots in a tract, and not as sections.

The descriptions should always carry the recital of the Ranch in which the area lies (e.g., Sec. 32 T. 3 S R 11 W in the Rancho Los Coyotes); the recital of the meridian (S.B.M.) is not strictly correct and in any event unnecessary.

Descriptions of Sectional type areas in Patented Ranchos should (and may safely) omit the expression "SBM" (or "MDM") if the Rancho is named, especially if the Section Subdivision is shown on a recorded map.

EXAMPLE: The NW¼ of Sec. 28, T 3 S, R 11 W in the Ro los Coyotes – – –.

A description covering a parcel of land lying in Sectional areas both without and within a Ranch grant must, for safety, recite each Section as a separate unit as:

> "Those portions of the SW¼ of Sec. 33 T 1 N R 9 W SBM, and of the SW¼ of Sec. 33 T 1 N R 9 W in the Sub. of the Ro San Jose as per map recorded in Book____ pg ____, Miscellaneous Records, described as a whole as follows, etc. – – –"

The section lines in the above recited case are *not* continuous and the fact must be definitely considered.

The location of corners must first be predicated on original (or replacement) corners set by the subdivider; if secondary or subdivisional monuments are missing, they must be located by prorate of the record measurements between found monuments; if no record be shown, assume Government sectional measurements, and subdivide by splitting for quarters, quarter quarters, etc. on the Government system; this splitting will apply to all sections, disregarding the rule of excess or shortage in the North and West Sections in the Township. 304

A description of a portion of a Rancho Section by fractional area (E ½, SW ¼ etc.) is construed as covering the area indicated by the normal method of subdivision of a section, restricted however, to the particular related dimensions of such portions as may be shown by the recorded survey of such section.

The fee title to land may be augmented or incumbered, or both, by easements; the augmenting easement is generally, but not always, connected with, but not lying within, the boundaries of the fee parcel; the incumbering easement always lies within the boundaries of the fee parcel, and/or the augmenting easement.

A granted easement may be appurtenant to the fee title which it serves or may be an easement "in gross" to the vestee of such fee title or to the vestee of some other title and not appurtenant to the land. Easements serving or associated with a fee title are considered as appurtenant in all cases where it is possible so to construe them; however, an easement granted specifically for operation of a public utility (pipe line, pole line, storm drain, etc.) will rarely be classed as an "appurtenant" easement.

Any appurtenant easement passes automatically with its parent fee, whether recited in the conveyance or not, unless specifically excepted; the easement in gross must be dealt with specifically. Even though generally unnecessary, the conveyance of an appurtenant easement is best established by a description of it as a part of the instrument.

Since most easements are of the strip type, the strip form, or in regular lots the directional form, is the usual style of description; irregular forms will be best shown by metes and bounds. It is always competent, and good practice, whenever the easement contacts the parent fee, to refer to such fee as a boundary; if the usual form, describing the fee as Parcel 1 and the easement as Parcel 2 in the same description, is used, then such reference can be made to ". . . the northerly line of Parcel 1 above described . . ." etc.

Always preface or include in the description of any area created as an easement, words definitely signifying the same, otherwise the *fee*, rather than the easement, will pass; for example, use "An easement (or right of way) over a strip of land _____ feet wide, the center line of which is described as follows. . . ."

The recital of grant of "a strip of land . . . to be used as a public street . . ." passes the fee, subject to restriction as to use; unless the statement is clearly made that an easement only is conveyed, it must be presumed that the grantor intended to pass the fee. A reservation, or incumbrance; for example—"reserving the East 30 feet for street purposes" is assumed to reserve an easement for a specific use; however, the proper wording is: "reserving an easement for road purposes over the East 30 feet."

If a deed grants a fee and following the description is a supplemental statement such as, "to be used for road purposes," you have

a dilemma. In such a case, you should "EXCEPT the interest in..." 400
and let the court decide what it is.

A complete understanding of the purpose of an easement is possible only if the uses which it is to assume are fully set out; a right of way for "ditch" need not imply the installation of "pipes" for carrying water in lieu of the ditch, but a right of way for "water service" would permit any installation and service compatible with local regulations. In any event the use of the easement must be stated; this form "an easement over a parcel (or strip) of land described as . . ." is uncertain and subject to much controversy as to the usability for any particular purpose, especially if the intent of use is not clearly indicated by the shape, position, connections or other matters in the description and instrument.

More than one type of easement may occupy the same ground, but the easement right and use by the senior title may not be interfered with by the junior, nor may the holder of the fee assert any adverse claim against the easement rights.

A grant of a right of way for a particular use without stating the width of the strip will be presumed to carry such width as necessary for access, enjoyment, and practical use of the right granted.

An easement does not become void, or non-existent if it ceases to be used for the purpose for which granted, unless the grant carries a limitation to that effect; nor can an easement be used for a purpose other than recited in the grant, without danger of attack.

An easement for "ingress and egress" is more limited in its use than one given "for street (or road) purposes"; this applies to both private and public grants.

An easement reserved "for street purposes together with the right to dedicate the same to public use," permits either the reserver of the easement or the holder of the fee or both to dedicate to public use, since a dedication by either does not prejudice the right of the other; any such easement so dedicated does not destroy the private right reserved; such right is subordinated to the public use, and is revived as a superior right upon vacation of the public easement.

Community easements affect similar strips of land adjoining each other, but on separate ownerships, and are described with defined easement rights, usually for driveway, walk way or utility purposes. The usual form of description is the strip or directional type. The easements may be created simultaneously or separately.

Often the owner of a parcel will deed a portion to another and create the community easement in the same instrument, such as—"Parcel 1. The South 50 feet of Lot 20 etc.; reserving an easement for community driveway to be used in common with a like ease-

400 ment hereafter described in Parcel 2, over the north 4 feet of said South 50 feet; Parcel 2; an easement for community driveway to be used in common with the easement reserved from Parcel 1 above, over the North 4 feet of the South 54 feet of said lot."

If the easement is to be created by itself, along a line between two adjoining owners, it may take the following form—"an easement for community driveway over a strip of land 8 feet wide, the center line of which is the dividing line between the lands of A and B, said line being the Northerly line of the land described in deed recorded in _____, and extending from X Street Westerly along said line, a distance of 375 feet"; of course the instrument must also include proper words of grant of the easement from each to the other, and be executed by both, in the similar form of any agreement.

Normally, the conveyance of a lot or parcel fronting on a street carries automatically with it, without recital in the deed, the underlying interest to the "center" of that portion of the street adjoining the lot. The underlying interest which the grantor may have or convey may, however, be more or less than half of the street. If the title line of the grantor stops at the side line of the street adjoining the lot line, then the interest in the street does not pass.

A vacation of a street does not disturb the existing fee title or private easements therein, but only removes the public easement. A public street held in fee by the governing body, subject to the street use, becomes a fee unincumbered, after vacation, but still in title of the governing body.

The transfer of title or rights from one government entity to another is accomplished in several ways. The California Highway Commission is authorized to relinquish roads outside of the main travelled way to the county or city wherein it resides.

Prior to September 20, 1963, it was uncertain if the county's fee in a road transferred to a city by annexation.

From this date to November 8, 1967, upon annexation of a territory to a city, county's right, title and interest in a road vested in the city.

After November 8, 1967, the transfer of title from the county applies to both incorporation of a city and annexation to a city.

The portion of a street *adjoining* a lot is that part "in front of" the lot and extending to the "center" of the street; the end lines therefore are lines drawn at right angles or radially to the side line of the street, through the corners of the lot, not the prolonged lot lines; the "center" is the thread or mid line of the street, but not necessarily (though usually) the center line as ordinarily understood. Because of possible interpretation of "adjoining street" as meaning the full width street, it is safer to recite the width taken, such as "the N 30 ft of M Street,

vacated"; "half" is frequently used, but is not as good form, although 400
"half width" would be acceptable. (Figs. 13, 16, 17.) Under certain circumstances or by court decision, the "right angle" end lines may be substituted by "prolonged side lines of lot."

A conveyance may be made to any portion of the street on which a lot fronts, between prolonged side lines, etc., but if such is the intention (and title) it must be so stated in the description.

A determination of "center" in many instances, especially on irregular width streets, is uncertain, and the particular parcel of land to which the fee of the street attaches or is associated is likewise indefinite in cases of dead ends or angles in streets. (Figs. 13, 16, 17.)

The right angling of end lines of lots into adjoining streets does not apply to those ends which lie on the prolonged side line of a parcel which line is also a boundary of the larger parcel out of which it came; in such case the attaching street area extends to the boundary line and prolongation.

The portions of two intersecting streets bounding a corner lot, which lie between the centers of each street produced to intersection, and the bounding lot, will carry with the lot, except in certain special cases.

Irregular widening, or angled streets with lot lines or corners at or near the angle points, or stub end streets, having lots abutting both on the sides and ends, produce uncertain conditions which make the attached street area to any such lots indeterminate in size or boundary; adjustment by agreement is the only logical solution. (Fig. 13.)

Easements dedicated on record maps for setbacks and for light and air, remain as an incumbrance on the land even though the streets adjoining them have been vacated, unless such easements are specifically vacated by the procedure outlined and authorized by legislation.

The vacation or destruction of an easement which is not dedicated to the public (*i.e.*, private easement), may only be accomplished by deed (usually quitclaim) from the easement holder to the owner of the fee encumbered by the easement. The object of the deed to-wit, to remove the easement and vest title in the grantee, is generally stated in the instrument. If the deed is to run to several grantees, the grant may be recited in the deed as—"in severalty to the various owners of land affected by said described easement, as their respective interests may appear."

Vacation of a public easement under the procedures provided by law for such matters vacates only public rights.

A general easement for any purpose (such as pipe lines, ditches, roadway, etc.) granted across a parcel of land, but not specifically

400 located, must be assumed to affect the whole parcel. However, actual and continued physical positioning of the easement in a particular location by reason of occupation by the objects of the same tends to fix such location as the intent of the grant, and if such fact is known, then the right of easement may be ignored as affecting any other part of the land; the exception to such assumption would be a clause in the grant such as, "easements for . . . now or *hereafter* to be located across said land," wherein it is evident that no portion of the land is free from the easement right.

A meander line of a stream or body of water when surveyed as such, is not a boundary line of title; it is only a line of identification or reference for the water line it meanders. (Michigan law carries an exception.)

Unless the contrary be shown, the boundary of an area adjoining the ocean extends to the line of ordinary high tide; adjoining a creek or non-navigable stream extends to the center or thread.

There are uncertainties and exceptions in some states, and the status of boundary on various kinds of water areas is different. Refer to Chapter V of Skelton's "Boundaries and Adjacent Properties" for a broad discussion of this.

A description calling for "the center of X Creek" as boundary is limited to the center as of the date of the instrument; the boundary does not move with a change in the creek alignment, but the contrary may be shown; there are some cases in which the court has determined the boundary to move with the change in alignment.

A described center line, by courses, of a creek or wash, cannot be accepted as a boundary line, if the creek is recited as a monument; it is acceptable as a meander only; however, the contrary may be shown.

If it is desired to grant to a line which follows the approximate center of a ditch or stream, use appropriate language to make the described line a superior call over the center of the ditch; such as: "thence along the following courses, being the approximate center of Carr ditch; South 26° West 470 feet, etc."; if the usual form . . . "thence along the center of Carr ditch South 26° West 470 feet, etc.", is used, the courses are meanders only and the grant carries to the center of the ditch, an indeterminate line. However, if *all* evidence of the ditch has vanished, and no possible means of location is available other than the courses of the meander, then manifestly, the meander must be assumed to control.

A tie to a water or water course boundary preferably should refer to the same as it existed (date) or as described (referred to) in deed (*record*); this does not apply to the ocean boundary.

A tie to the ocean may state . . . "to the line of ordinary high tide of the Pacific Ocean; thence along said line (or said Ocean). . ."

Any description of land bounded by the ocean should carry the exception of "any portion not formed by the natural deposit of alluvion in imperceptible degrees" or similar language because artificial accretion is held to belong to the State and the line of ordinary high

500 tide as a boundary must be the line at which the tide would be limited by natural accretion. Ordinarily the line between natural and artificial accretion is indeterminate.

Mean high tide, ordinary high tide, and the line locatable at the elevation of mean tide established by the U.S.C. & G.S. are usually considered synonymous; a survey of the line of ordinary high tide at any particular date, then, will be a location of a traverse line of a contour of the known elevation at that date. The Federal rule for the position of mean high tide is the average of all mean high tides over a period of 18.6 years, a lunar cycle. Some states have different rules and many conditions await court decisions. Due to constant change in coast lines, any survey picture is good only for the moment for which it was made in most all cases.

Whereas the survey lines of Federal lands bounded by ocean, lakes, swamps, etc., are meander lines, the survey lines used for describing the patent to State Tide Land Locations are not so considered. Thus far the latter has been considered as a fixed line with the possible exception of any priority of a Federal grant along the same boundary.

The extension of side lines of water front property to the line of tide as it recedes due to accretion follows the same general rules as in vacations, at right angles to the tide line; this is technically an indeterminate condition as the direction of tide line changes at each instant. Ordinarily, the prolongation of the side lines may be acceptable, if approximately perpendicular to the water line, but the contrary may be shown. The only safe way is agreement betwen adjoining owners fixing the direction of the extension by definite land ties.

No attempt should be made in descriptions of land fronting on ocean or tide water boundary to add interests or claims seaward of the tide line, if the description recites the line of ordinary high tide as a boundary; it is dangerous because such interest or claim which actually carries with the upland is automatically included in the grant and no other claim is valid except by a special grant; any attempt to specify or limit the interests involved may result in an unintentional separation of some rights which would otherwise pass.

Any description calling for distances "to the line of ordinary high tide" should state them as "more or less" or refer to the line as "established by survey on (date)."

The surveys and descriptions of mining claims for patent are governed by definite regulations of the Federal Government; the Manual of Instructions for the Survey of Public Lands gives the procedure.

Any Licensed Surveyor or Registered Civil Engineer may make surveys of claims, except for patent. Patent surveys must be made by a U. S. Mineral Surveyor.

For all practical purposes, except for U.S. Patent, wherein the form is rigid, any good form of description for a mining location is usable. The essentials as to ties, recitals, metes and bounds (exclusively), with all bearings and distances, must be used.

The weak and uncertain forms used by locators in filing claim notices in the Recorder's office or in the Mineral District office are to be deplored; however, they must be accepted, and their physical position determined if possible, by survey, before attempting to write a proper description for any documentary purpose.

700 SURVEY LOCATION

701 *Relocation from maps*

The general theory of relocation of any lot or portion of a subdivision map is the re-establishment of the boundaries and corners of the lot consistently or proportionately to any and all other portions of the subdivision record; this may involve field work of considerable extent, recovering necessary existing monuments, street center lines and long-time possession lines all within the subdivision, and any associated monuments or lines adjoining outside the subdivision, especially those which may be shown on the record map of the instant property as the lines or boundaries.

Inasmuch as all matters shown on a map of record are competent as data affecting the location of any part of the map, the relocation survey must of necessity take cognizance of them in the relocation survey, as far as they may bear upon the particular lot or line in question.

Probable original errors or uncertainties in boundary lines of recorded subdivisions, especially in the boundary street positions of maps recorded in Miscellaneous Record books (old subdivisions recorded prior to 1900) give preference to interior street lines of the subdivision as a better basis of relocation of lots.

Any considerable difference between record and existing distance from interior to boundary street is probably not proratable, and the record distances of lot lines, carried from the interior street, is indicated, unless a definite series of occupations of long standing shows otherwise.

Location of lines and boundaries on a map which is definitely inconsistent with the boundary maps or relocated exterior street center lines should be made from the data and monuments of the map itself, using, generally, the record distances shown on the map, letting the surpluses and shortages fall on the boundaries, regardless of conflict.

It is usually more accurate to assume the original survey was incorrect in its boundary location than in its interior.

Examples of error in boundary street information on maps of areas in Los Angeles City:

(1) Mott Tract between Court and Temple; Temple Street as located in the tract is too far north, producing shortage between the present streets; proper location of lot lines is record from Court Street, in spite of condition that many holdings are on the Tract record of lot distances measured southerly from the existing line of

Temple Street; good title here is only obtainable by deeds or court actions; assessments, deed descriptions and record maps are frequently at variance with each other and to the land occupied.

(2) Several tracts, particularly in Hancock's Survey of Los Angeles City land and Temple and Gibson tract, where it is definitely known that a fairly regular surplus (approximately 0.7 per 100) exists; resubdivisions in these areas often show the record dimensions of the original; necessary here to determine correct positions from *interior* lot or street lines of long standing and work outward by record distances or prorates to other interior lines.

Relocation of curved lines on maps, especially curves of street boundaries needs a complete analysis of the probable original layout. Since arc lengths are invariably computed, they may nearly always be ignored except as a check. The most common difference in resurvey shows in the curve delta. If the radius is a fixed whole number (300; 750; 4°; etc.) it is probable that such is an assumed figure originally, and should remain constant in relocation. A fractional radius (45.22; 654.3) is usually computed to fit tangents or other curves and may be varied to fit adjoining conditions. The curve tangent (semi tangent), if a fixed whole number, should be considered similarly to the whole number radius; it is rare to have both radius and tangent whole numbers. If both radius and tangent and adjoining curves are odd, a prorate adjustment between fixed points on adjacent courses or to discovered monuments is indicated. Except in rare instances, broken back curve points or creation of tangent lengths between compound or reverse curves is not warranted in relocation. Do not use station numbers and pluses for re-run purposes except as information, unless they are definitely the basis of the location, and their ground position is determinable without question. Run out the tangents to the curves to a P.I., and back up the calculated T. to B.C. & E.C., based on the given radius and measured delta. Railroad track curves and right-of-way posts are unsafe and uncertain monuments; likewise do not depend on the California State Highway concrete "C" monuments as accurate, if set prior to about 1935; if the uncertain monuments have to be used in the absence of anything else, get enough of them to obtain a fair average adjustment which will correspond closely with record.

If both the P.I. and B.C. and/or E.C. monuments on a curve are recovered and together will not fit the record tangent or radius or both, then additional information must be obtained and the most probable relocation made, keeping fixed, either the radius or tangent length as above analyzed, and ignore the located monuments as

701 defining the indicated B.C., P.I., or E.C. except possibly the alignment of lines extending beyond the curve. The frequent practice of some engineers in accepting supposedly correct monuments and changing fixed radii or semi tangents to produce approximate agreement is definitely poor if not inaccurate procedure insofar as it is an attempt to re-establish the *record.*

Do not depend on relocated B.C. or E.C. or flat or sharp angle points as an origin from which to measure record distance to a nearby lot corner; always establish the lot corner by its prorate from nearest adjoining corners, or streets on either side or, if indicated, by record from a nearby corner of a block or lot. The City location of curves, with changed radii or other data, is acceptable as a tie only, and may or may not be usable for lot or curve relocation. The city lines are usable as defining the recognized street boundaries, but such boundaries may be created by occupation or other prescriptive method. Example, the Encino Tract, in which the street work was erroneously built about six feet off record position; the Borden Tract in which the city fixed a 60-foot street when the monuments clearly indicate a greater and variable width; other tracts which can be accurately relocated, in which the city center lines have been changed one or more times and which did not or do not now conform with the original position of record or existing original monuments of the tract. The above also applies to County, State Highway or Flood Control lines, whether curved or straight.

702 *Location of Lot Lines in Blocks*

If the relocation of the exterior lines of the tract or blocks therein shows substantial agreement with record, both as to angles and distances, the prorate method of fixing lot lines is indicated; excepting in those cases where a very definite group, or the majority of the holdings, show a consistent shift or variation in one direction, and the age of the holdings indicates their location from original monuments.

The City Boundary Tract is an exception to the above. Its boundaries are correct, based on previous survey, but the lot lines and interiors were not originally monumented; a later survey made erroneous location throughout the tract and possession followed the error. In this instance correct location is made from the original tract layout as of record. The titles must in each case be adjusted by proper methods, changing the record to fit the occupation or vice versa.

If one or more of the block exteriors is shown to be incorrect or uncertain, then preferable to lay lots at record distance from the certain lines, throwing all variance in the uncertain lines. 702

If the uncertain exteriors are at opposite ends of the block, then best to locate record both ways from a reasonable determined group of holdings in the central part of the block, giving the end lots the differences; this group of holdings used as a base.

If the block is vacant or irregularly occupied and the exteriors are uncertain, no relocation is safe; a new record map, or an arbitrary location with appropriate agreement deeds is required; Walteria south of Torrance, is an excellent example; any work done in this area should be very carefully reviewed.

If a record map shows no bearings, and insufficient distances to determine direction or position of lot lines, the occupation of long standing will have control preference in spite of possible apparent differences from record. The best and most satisfactory solution here is a replatting and agreement deeds.

It is permissible to locate on a prorate basis in one direction and record in the cross direction of the block, if the evidence so indicates.

No street cut through a block by deed or otherwise, even though the deed recites a lot line as the street boundary or gives a definite tie to a lot line, subsequent to the original record, is usable as a base for lot relocation if opposed by *any other* reasonable positioning of the lots. The location survey must be made of the whole block, as though the later street did not exist. This situation is equally true in any attempt to locate original lot lines by use of boundary of resubdivision covering part of the block.

A series of regular lots with a single odd lot on the end indicates the location by record, of the regular lots, with the difference in the end lot; this is not a safe method for relocation unless the block exteriors are materially in error; the prorate method is advisable; odd lots at each end of the block indicate similar treatment.

A block containing some lots with no dimensions will in general carry *record* for all dimensioned lots, with all differences applied to the undimensioned lots.

The proper location of *any* lot in a block must of necessity include a study and sufficient measurements to verify position in the *whole* block.

Use of record angles for side line position will seldom prevail over the distance location on the end lines; the complete analysis only, will determine the preference of such use.

It is not competent to assume a line fixed by agreement as a lot line or division line between holdings, as a correct lot line basis for relocation of other lot lines in the block; as stated before, an original

702 lot line never changes, and an agreed line, per se, cannot become a lot line for purposes of determining other lines in the vicinity.

In every case of property line determination, whether physical or record, it is absolutely necessary, for correct title interpretation, to assemble all the facts, record and survey, including possession control, concerning the property and its boundaries; with this data, set up all the probable or possible theories of position, considering fully the physical, engineering, title and legal factors and their interrelation; ascertain a solution which will fit the majority of the analyses most closely. This is the theory of *majority probability* and normally will produce the most satisfactory results. There are certain cases in which other factors will take the place of majority probability, example: the fixing of a definite boundary line by agreement and deed; or decree of court determining a boundary by testimony and evidence or other matters stranger to the title or survey and not of record.

703 *Survey Location of Described Lands*

It is not much of a problem from the conveyance standpoint, to make an original survey of a parcel out of a larger piece, the only item other than care and accuracy in the work being the correct location of the boundary and corners of the larger parcel affecting the new work. Likewise, the description from such a survey is comparatively simple, and follows the basic rules with little need of involved interpretation.

Relocation survey of a parcel (which today accounts for most of the surveyors' work), on the other hand, from a description of record, associated with monuments on the ground which must be verified as acceptable for the purpose of the relocation, raises many questions, and often presents difficult or possibly unsolvable conditions. The surveyor who attempts a proper relocation must have a fine knowledge of title as well as engineering interpretation, discrimination in matters of choice between factors, and good judgment in assigning proper weights to the various information, both record and physical.

No relocation should be attempted without first assembling *all* the pertinent record data and analyzing it by computation, plat, and other means; then make the survey, tieing in *all* monuments, occupations and other matters which may have a bearing on the boundaries of the land in question; next, combine the record and the survey, re-analyze and determine the position of the corners which will most correctly satisfy the conditions. Only then can the property be safely monumented.

There are certain occasions wherein a survey should fix boundaries *inside* the possible limits of a location, to prevent uncertainty, or to locate a line at its *minimum* extension toward an adjoiner rather than its maximum, when the analysis for either position is of equal weight. Such determinations should be stated to the land owners with a recommendation of adjustment to modernize and fix the boundaries with certainty. Such treatment of location will leave gaps between adjoining lands rather than overlaps, in most instances. **703**

Uncertainty of record description and/or uncertainty or indeterminability of physical monuments may make a location impossible; in such an event, the only solution is establishment of a substitute line, or lines, by agreement deed, with mutual quitclaim, or by court action. In many cases the agreement is a more satisfactory method for fixing a line than all the labor of analysis and extended survey, even though the relocation is reasonably possible.

It is essential in making a relocation, or in describing a parcel of land, to have all the necessary record information concerning it. Off-record data is also helpful and on occasion may be vital, if the record is indeterminate.

All maps, field notes of survey, deeds and other documents pertaining to a location should be gathered together and completely analyzed before completing the work.

This Section deals primarily with definition and relocation of corners of surveyed land under the U.S. Rectangular System of Survey, commonly known as Sectional land.

The statements and analyses are given for the use of private surveyors as distinguished from those doing Government work under control of the Bureau of Land Management, wherein certain procedures of somewhat different nature prevail. The private surveyor in relocation of section boundaries must, naturally, conform as closely as possible to the rules of the Manual issued by the Bureau, but, as stated in the Manual, when sectional land passes into the hands of private persons, the responsibility of the Bureau ceases, as to locations, except in an advisory capacity; the private surveyor must supplement and interpret his relocation problems with the best and legally conformable practice of surveying in his State or area, relying on the Manual for those procedures or rules which are not controverted by such other practice.

The definitions and methods herein given are applicable to any type of relocation, if the data are available, regardless of the land being "sectional" or otherwise.

Existing Corner

An existing corner, as the term implies, is an original corner monument in place.

If described in a document, or if clearly identified and associated therewith, the monument becomes a record.

To be a proper record, and usable as the true record corner, the monument must be proven and identified as the original (or part thereof) and in its original position. Age, impermanence of material, disturbance of surroundings, often make reliable information difficult, and thorough investigation must be made to substantiate the position.

When so identified, the monument becomes a superior call in description and location.

Existing Corner Position

An existing corner position is a position of an original corner or corner monument (now non-existent), locatable either by provable data, or by a known relation between the original corner and a replacement (or witness) monument.

Relocation requires full investigation and verification of the position, including such items as agreement in distance or ties to other known and acceptable corners, testimony of persons who knew the original monument and/or position, documents showing relation between the original and present position (if different), etc. **704**

If a monument replacement is involved and the evidence indicates an offset or witness type position of such replacement, with a specific relation by dimension, to the original position, then such replacement monument, when verified, can be accepted as a proper locative of the original position, using such specific dimensions.

Under this grouping falls the relocation of a corner position by means of bearing trees and witness corners as recited in Government Field Notes of Township and Sectional surveys; also by ties in the Field Notes of public and private surveyors.

A corner position may be recovered from given dimensions (usually of limited extent) to natural or artificial monuments noted or recited in documents of record, including maps showing such information; provided such monuments are verifiable as to their own position.

Obliterated Corner

This type of corner is one in which all traces of the original monument, or its immediate accessories (mound, pits, etc.) have disappeared, but whose position can be recovered by testimony of witnesses, acceptable record evidence, or measurements and other data in the field notes (or available maps) relatable to such monument.

Although occasionally found in other survey locations the obliterated corner is generally associated with sectional surveys. The principal reason for this is that this corner depends almost wholly for position on the topographic record of its location; such a record is most frequently found in the Field Notes of the Sectional or Township surveys.

"Obliterated Corner" (or Monument) and "Existing Corner Position" are often considered and used as synonymous designations.

Location of the corner (or monument position) depends on the field note (or map) record of the survey setting the corners, showing dimensions between the monument, its accessories (bearing trees, witness corners, etc.), and/or topographic calls along the lines of survey; also may be considered items noted under Existing Corner Positions, above.

The topographic assistance of the field notes is especially important in locating obliterated corners. Competent showing of topography, particularly if near the corner in question, enables a very close and

704 substantial replacement of the corner, even when the dimensions are not as conformable as they should be.

1. Locative Topography.

A call for a sharply defined topographic item, such as crest of a steep bank, crossing of stream of narrow persistent channel, foot of abrupt rise, and the like, particularly when within short distances from the corner, provides an excellent locative position, the corner being established at the stated distances from the nearest or most precisely locatable items. If there is a discrepancy between distances from the corner to two or more of such items, the choice of use will depend on preference, or conformity with other factors, such as reasonable agreement with artificial objects, majority conformity with other topographic calls, etc.

2. Confirmatory Topography.

Topographic features which are variable in position, as a movable stream flow line but with a broad bed, or indefinite as to precise position, as the top of a hill which is flat topped with considerable distance from brow to brow, or the base or spot on a sloping hillside where the field notes state "enter valley," or indefinite as to choice between two or more similar items where the measurement call in the notes actually falls in between the items, or too great a distance from the corner in question to place reliance on the measurement where a closer feature is more locative, all are rated in the category of confirmatory topography.

Although such topography may be unusable for location, it nevertheless furnishes good check and substantiation to the position determined by locative calls. Occasionally a confirmatory position, such as a small stream crossing or top of a sharp rise, even at considerable distance from the center, may also be used as a locative call. If the notes carry good and sufficient topographic calls, the lateral position of the lines of survey may be determined as closely as the desired corner.

Likewise, any shift in line or corner location through topographic examination may discover the traces of original corner, even though considerably out of record dimensional position, thereby eliminating use or reliance on prorate methods.

If, as occasionally occurs, an obliterated corner, which is located by good but possibly not "pinpoint" topographic calls, can also be positioned by prorate between two proven corners within a small difference (say up to 10 feet) between the locations, it may be advisable to accept the prorate position as a "substantial" location, and

consider the topographic location as confirmatory. This method, 704
however, should be used sparingly and judiciously, since prorate is held to be the method of *last resort.*

California courts have consistently held to topographic methods of relocation where possible, even in cases of wide discrepancies in dimension, when the field notes state specific relations between the corners and existing natural features, fixing the corner position as nearly in conformity with dimensions as the recited and existing topography permits. The approved Field Notes are held unimpeachable as to statements therein, such as topography, description of monuments set, etc., but recognize the possibility of discrepancies or error in dimension.

Lost corner

A lost corner is one whose position is indeterminate from original traces or other reliable evidence such as mentioned hereinabove, and whose relocation is possible only through a system of relative measurements from associated existing corners or corner positions. "The engineer is *not prepared* to consider the restoration of a *lost corner* until he has *exhausted every other means of identifying the original position*." (Italics added.) See §361 of the Manual.

Lost corner positions are determined by the prorate method, between existent corners (or corner positions) on each side of the lost corner.

For Sectional land, the procedure is laid down in the Manual of Instructions for the Survey of Public Lands, issued by the Bureau of Land Management of the U.S. (formerly the General Land Office). For this procedure see §s 360 to 385 of the Manual. The methods described for such relocation are to be followed for lost corner restoration unless other considerations, or laws and practices of the State intervene.

Relocation of lost corners on privately owned lands, not within the scope of the Government system of Rectangular Surveys, is determined by several methods and analyses set out or provided elsewhere in this Syllabus, or by the use of methods of local custom and sound surveying practice.

The locatability and ground position of boundaries of a description, or the question of actual title held under a description, requires a true analysis and interpretation of all the factors involved, and statements made, both in the instant property described and the surrounding properties thereto; all lines and points already of record must be precisely established, based on the methods and directions originally employed to fix them, and properly associated with the boundaries of the land in question; the matters of age and position of occupation, and of common practice and consent in the locale, must be considered; mathematical accuracy, superiority of call or tie, sufficiency of the description after casting out proven errors or repugnancies and uncertainties, must all be taken into account. The resultant determination, balancing all factors and accepting the greatest number (properly weighted as to value) which are generally conformable, forms the theory of majority probability.

Although the final decision as to title of a parcel, or any portion of it, rests with the courts, it is necessary for the title engineer, surveyor, or land owner, to possess sufficient knowledge of descriptions to determine with reasonable assurance that the written evidence of the boundaries is a true representation of the physical location already established, or to be established, free from conflict with adjoining possession or title. If such conflict is inevitable, then the most feasible conformity must be determined by the evidence of the majority of factors which agree or support one another.

801 The necessity of resorting to court interpretation of intent or title may be largely avoided by the use of descriptions which are constructed in words and terms that are unequivocal, exact, and bear the approval of legal precedent, mathematical definition, and engineering terminology. In those instances where a difference arises between legal and engineering interpretation of terms, methods, or phrases, the description should be so worded that its validity or certainty is unquestionable, regardless of the direction of approach in analysis.

Any description drawn from data furnished by a survey must rest on the assumption that such data are correct, unless the contrary be shown by check against the records and/or mathematics, in which cases such data must be revised.

On the contrary, any location or relocation of a description on the ground or to adjoining described areas, must be interpreted as to its sufficiency and agreement with existing conditions and record, by

legal title, and engineering analysis which will show intent, and facts 801
of agreement or conflict, the nature of the tenement conveyed, and the incumbrances imposed.

Every instrument must be read from its "four corners" to ascertain intent. If the intent is not discoverable, parol evidence must be introduced and considered, to prevent voidability for lack of certainty; if the intent is apparent, but the description is impossible, due to error or multiple possibility of location, there is grave probability of voidable status.

Any errors or discrepancies which disturb the normal or sensible 802
interpretation, and which by omission do not alter the sense or accuracy of the remainder, are to be cast out; any errors evidently clerical (such as "Ely" instead of, correctly, "Wly") which by alteration will give a good description, but otherwise produce an absurdity or impossible construction, may be assumed as corrected in making a full interpretation; however, such assumptions must be taken with *great caution*, for it is possible to have a description which can be fully construed either with or without the change.

A *grant* deed conveys to the grantee the fee title held by the grantor of the area within the described boundaries, together with any appurtenant rights, subject to any reservations or exceptions named in the instrument, or of record. If there is any portion of the land described which is not in title of the grantor, it does not pass by the instant grant but will automatically pass by any subsequent acquisition of the same by the grantor, as "after-acquired title" under a grant deed passes to the grantee thereof.

A *quitclaim* deed conveys "all interest," only, of the grantor, whatever it may be, in the land covered by the description, and does not carry any after-acquired title. Quitclaim deeds are generally used to clear easements, reservations, uncertain rights, undisclosed or nonrecord interests, etc.

An *agreement* deed (so called) for fixing lines between adjoining owners should contain, besides the terms of the agreement (including proper description), words of conveyance to fix the title on each side of the lines, otherwise an estoppel only may exist, which will possibly not extend beyond the activities or claims of the parties to the agreement.

An *exception* in a description withholds or keeps out the excepted portion from the conveyance.

A *reservation* in a description is a conveyance of the whole with an immediate taking back by the grantor of the portion or tenement reserved.

802 Many times in older descriptions, the words "excepting" and "reserving" were used together or synonymously notwithstanding the fact that they are contradictory in meaning. "Exception" is generally used with reference to the fee, while "reservation" is usually attached to an incumbrance, right, or easement in or on the fee.

An *easement* is a right created for the benefit of others than the fee owner of the land on which it exists; it is known as the servient tenement, the fee being designated the dominant tenement.

In order for the "exception" or "reservation" of a strip of land over a granted fee title, to be that of an "easement," the words of easement or right of way should be clearly stated, otherwise, both the fee and easement in the strip may be retained by the grantor. The exception of the strip, "to be used as a street" or "for road purposes" or other such language does not create an easement, but merely restricts or designates the use to which the strip may be placed; under certain circumstances and by presented evidence, the courts may decide the contrary.

803 Many recorded maps show, and may state by note thereon, that the areas and/or distances shown on the map extend to the center (or other designated lines) of adjoining streets. It has been presumed by many, and the *undisclosed intent* of the grantors has been, to assume the lot lines and corners on such maps to be along or on the dimensioned lines. Except in certain special cases where evidence to the contrary is preponderant or where the map of record clearly states the fact that the lot lines are to be taken as the street center lines it should be assumed that the *lot lines* are the *side lines* of the streets, regardless of the location of the distances on the map; that is, the lots on the map are to be considered as those areas bounded by the solid lines, and do not include the streets shown thereon, although the fee title to the center of the street (if in the grantor's title) carries with the fee of the lot adjoining, whether recited or not. In doubtful cases, a full analysis should be made, considering all factors, as possession, neighborhood claim and assumption, consistency of holding in the tract, assessment, legal interpretation, etc.; any description in such area should definitely state the dimensional position with reference to the doubtful line, such as: "West 2 acres, measured from the center line of M Street," or, "the east 300 feet, measured from the west line of Y Street."

If it is desired to describe an area on the assumption of street center lot lines, the following forms are acceptable as indicating the *intent* (not necessarily the fact) of the call:

(a) The East 5 acres of Lot___, measured to the centers of adjoining streets.

(b) The North 330 feet of Lot___, measured to the center of "X" Street. 803

(c) Metes and bounds, reciting for instance: "beginning at the intersection of the center lines of "X" and "Y" Streets at the Northwest corner of Lot___; thence along the center line of "Y" Street, being the North line of said Lot, etc."

The best form, if it is desired to give street center dimensions, is a metes and bounds description along such lines, prefaced by this phrase: "That portion of Lot___ included within the following boundaries . . ."

Location of property on street center measurement maps with the deed in the form: "N half of Lot___," or "E 5 acres of Lot___" (record 10 ac) is difficult and often indeterminate. The assessors (in Los Angeles County and cities therein) usually, but not always, assess such land on a street center basis; Torrens titles accept both the net and gross theory; occupation may be on either basis, frequently on each within the same original lot. The safe plan on the part of the surveyor is to mark the lines on a gross (street center) basis on property fronting the streets, and on the *lesser* area (or distance) on the areas adjoining the *blank* lot lines, and on the *least* area (or distance) after giving the adjoiners the greatest area (or distance) on parcels in the central part of the lot. *Then tell the client* the circumstances of the case and recommend an adjustment to acceptable lines between his and adjoining holdings. Occasionally long standing occupation will make the location definite and the deed may be safely construed by the surveyor as covering the occupation and the occupation may be safely insured by a title company.

A line between two fixed designated points is a straight line, unless otherwise stated; the use of "in a direct line" or, "in a straight line" is unnecessary except for clarity or distinction as against another line which is curved or angled.

Distances between points or lines which are not otherwise defined 804
or limited are fixed and exact; no proration of such distances is permissible.

Distances given as "more or less" between definitely defined points give way to the actual measurement between those points, but if the points are not so defined, then the "more or less" is meaningless, and the distance is fixed.

The use of "more or less" except as explanatory, is superfluous, as the nature of the termini of the dimension will control.

Remember that a lot line shown of record in a certain position, or 805

805 a lot shown of certain dimensions *never changes* as regards the record. The record may be in error and a new physical location or dimension may be created which is acceptable, but the exclusion of a portion of the lot for street purposes or other reasons does not reduce the lot to its new usable area; any description reciting a portion of a lot with the idea of defining certain dimensions within the usable portion *must* consider the *whole lot less the excluded part.*

The analysis of a location problem may be a matter of relationship of dimensions, relative dates of conveyance, consistency between units of a group of physical monuments, legal interpretation of terms of the descriptions, or other related items, or any combination. Every problem calls for its own particular solution and no two are exactly alike, even though the premises appear identical.

Reference to "the NE corner of the land *conveyed* in deed to J. Jones . . ." may be an entirely different corner than "The NE corner of the land described in deed to J. Jones . . ."; a lot may carry the fee title to half the street adjoining, so that Jones' *conveyed* corner may be in the center of the street, while the *described* corner is in the side line; also the description may include other lands together with that conveyed (very common in blanket form deeds); make certain that your description refers properly to the proper corner or line of the land "described" or "conveyed" as the case may be.

Any reference to a deed which contains more than one description properly is made by reciting the particular parcel in such deed to which reference is made, *e.g.*; ". . . to the NE corner of the land described in Parcel 2 of deed to John Doe recorded . . ."; otherwise the description may be uncertain; it may be possible to make a perfectly describable parcel by recital to any of several parcels of a reference deed.

In general, reference to land "described in deed" must be taken to refer to the net result of such description; that is, the whole description *together with its exceptions*; if any uncertainty is created by the general reference because of such exceptions, alter the language of the reference to remove the doubt.

Contrariwise, reference to land "conveyed" must be analyzed to determine if there is more area or less area conveyed than that actually described. (For one example, see Section 400, page 70.) If the description covers more area than title is vested in the owner, then the conveyance is less than the area described.

Any document or simultaneously recorded set of documents, in the nature of a partition, having a common grantor, in which the intent is clear to divide a parcel into several mutually interdependent parcels, with common boundary lines, must be considered, as to description, of the same character as a block of lots on a record map; that is, in any resurvey or relocation, the dimensions of each parcel must be prorated to the total of the including parcel, in similar fashion as lot frontages prorated to the total frontage of the block. Subsequent descriptions of any of such parcels should be tied to the adjoining boundary by recital of the proper deed reference, thereby keeping alive the fact of proratable instead of fixed dimension. **805**

Tie distances and/or bearings from monuments exterior to the boundaries of a description will control the boundaries unless priority of call or existing monuments within the described area (such as railroad tracks or utility lines, recited or defined in the description as monuments), or language in the document, subordinates the exterior monuments to the main boundary of the description. The word "monument" here, refers to either physical or record or both. **806**

In spite of the general holding of the courts that ground monuments govern over distances or other calls in a description, the contrary may be shown, especially if it be shown that the monuments were set for and subsequent to the description and conveyance and the court will not ignore the showing of intent as opposed to such monuments. The physical location of a parcel of land may precede or succeed the record description of it; if it precedes, then clearly the monument will hold over the description, unless shown to be disturbed or definitely in error, while on the other hand, if the location succeeds the description, it should be assumed that variance of the ground location is due to error in the survey or in the interpretation of the description. Also, if a physical location of a corner or line is assumed to define the record call thereof without actual mention of it in the record, such location may be shown in error.

The record monument is presumed to be unchangeable as compared to the limits of permanency of physical monuments.

As long as a recited physical monument exists undisturbed (such as a wall, tree, concrete post, ditch, etc.) it will define the boundary of the described area, but after the destruction of it, the boundary relocation must depend on the dimensions of the description, associated with any others in adjoining titles which may assist in determining the true position.

The dimensions in any survey or description are presumed to be exact and true, unless qualified or provable to be in error in some particular; this theory gives rise to the prorate system of remeasurement.

As a general proposition, proration is usable for analysis purposes but not as a consistent or precedent method of location if opposed by other factors more closely conformable to title; the contrary may be shown, however.

806 A metes and bounds description without boundary recitals, and showing a mathematical error in closure has the whole error, legally, placed in the last course "to the point of beginning"; the usual engineering interpretation, however, will place the error in that course or courses which most nearly conform with the bearing of the calculated course of error closure; if additional information is available, then the adjustment may be of a different nature.

A description reciting "thence Ely ____ ft to the West line of land of M . . ." where the distance given will not carry to the line of said land, will be presumed to carry to said "West line" along the prolongation of the described line, but the contrary may be shown.

A described parcel of land carrying an exception of a part (or "any part") is not affected or limited by the exception if such part it not actually included; the form, "except any portion included within . . ." or "except that portion if any included within" is often added to a description to exclude possible strips not in title, whose position is indeterminate, unknown, or suspected, but which may not in fact affect the described area.

807 You sometimes have multiple possible combinations of physical and record solutions to the determination of a line or point to be shown on a map and used in a subsequent revision of a description. The decision as to which one or combination to use is not based on one rule only; the decision should be the result of careful analysis of both the singular facts and the different combinations of them together.

When one grouping of a set of facts conforms more closely than any other combination to the conditions set forth by the description on record *and* the original survey, you then have the most reasonable solution that is probably available. This is known as the theory of majority probability, and would likely carry the most weight in court.

The correct construction and interpretation of a description depend on the proper use of the words and phrases therein and their position within its structure.

Many parts of passable and interpretable descriptions are rhetorically or grammatically incorrect, but by reason of common usage possess a definite meaning. Wherever possible, it is best to use the correct language rather than the common form or idiomatic expression:

FOR EXAMPLE (1) "to a point in a line distant thereon" is incorrect, although frequently used, for "thereon" grammatically refers to the "point" not the "line."

(2) "as per map (or deed)" is in general use and is acceptable as satisfactory and possibly preferable to the synonymous phrase "in accordance with."

1 Adjoining; 2 contiguous; 3 coincident; 4 adjacent

1, 2 and 3 have common boundary in part.

4 applies to items in the same vicinity.

2 has contradictory definition in Black's Law Dictionary, do not use this term.

1 and 3 may also be 4 but the contrary is not true.

3 applies to two identical boundaries or lines.

4 qualified as "immediately adjacent" becomes synonymous with 1 or 2.

1 applied to the portion of a street area contiguous to a parcel of land means that part of the street "in front of" the parcel, lying between the parcel boundary and the "center" of the street, or the line of the vacated portion of the street contiguous with the parcel; may apply to the whole width of the street, so when used, the definite line in the street should be stated such as "the North 30 feet of the street adjoining," or "all of the 60-foot street adjoining" or "that portion of the street vacated (if less than half is vacated) adjoining" etc.

Along

Referred to a line carries recital on and in the direction of the line and occasionally on its prolongation.

Referred to a street or strip or way or area having width, without

900 qualification, carries the alignment of the center, or thread of the street or strip.

1 Along a line; 2 from point to point

1 carries the dimension on the recited line through its various changes of direction.

2 is a dimension on a straight line; unless qualified, a tie or distance between points in the shortest measurable length, *i.e.*, in a direct line.

1 And; 2 and of

1 Normally separates recited parts, including the exception and/or portion attached to either part; but it may be insufficient to establish clearly the application of the exception.

EXAMPLE: "Lot A and Lot B, except the E. 50 ft." might be construed to except the 50 feet from both lots or only from Lot B.

Specifically, use

"Lot B, except the E. 50 ft. and Lot A" or "Lots A and B, except the E. 50 ft. of Lot B"

for the exception from one lot;

and "Lots A and B, except the E. 50 ft. of said Lots" or "Lot A except the E. 50 ft.; and Lot B except the E. 50 ft."

for the exception from both lots

2 "and" when succeeded by the word "of" establishes the equality in application of the precedent or succeeding part stated, to both parts.

(a) "The E. 50 ft. of Lot A and of Lot B," is the E. 50 ft. of each of the lots

(b) "Those portions of Tract X and of the vacated street, described as (a whole as) follows . . ."
The parenthesis is optional, though usually desirable.

Plurality in similarly designed areas connected by "and"

"The E. 50 ft. of Parcels A and B"; or "Blocks A and B, except the the E. 100 ft. thereof";

carry the equality of application of the portion or exception to each of the parts named.

For example: ". . . point is located in position by the bearing and/ or distance. . . ." Indicates the probability of application of both the precedent or subsequent terms, or of either; the determination rests on the interpretation of the context.

Being

Used as supplemental or in addition to previous identification; frequently to distinguish between superior and secondary calls.

Beginning; Commencing; Point of Beginning; Place of Beginning; True Point of Beginning

Descriptions "begin" or "commence" at a point and close to the "point of beginning" or "point of commencement"; or they begin at a reference point, continue through several courses to the "true point of beginning" and from thence continue around the boundaries of the actually described area closing to the "true point of beginning." The distinction between the initial point of reference and the initial point of the bounded area is sometimes made by using "commencing" for one and "beginning" for the other.

The usual form "beginning at a point . . ." should complete the description with ". . . to the point of beginning."

The form "beginning in the line . . ." should complete the description with ". . . to beginning" or ". . . to the place of beginning."

1 by; 2 on; 3 with; 4 along

1, 2 & 3 are old forms for 4 and not as satisfactory; better to use "*along* a line," rather than "*by* a line" or "*on* a line," or "*with* a line."

1 center; 2 center line; 3 transit line; 4 line of street or way

1 applied to an area, it is the center of equilibrium of the closed figure.

Applied to a strip (*i.e.*, a line-like area, having width) it is the "thread" or line midway between the side lines of the strip.

1 applied to a curve, generally meant as the midpoint of the described arc, but may be the center of the circle of which the arc is a portion of the circumference; when used with curves, the definite statement should be made as "the center of the arc of said curve" or

900 "the center of the circle of which said curve is an arc," as the case may be. (See *circle* etc., post.)

2 in a street, without qualification, recital of the "center line" is generally taken to mean that of the original street, before whole or partial widening; if widened throughout its length to a new standard width, wholly on one side; "center line" may possibly be taken as that of the strip as widened. Any such ambiguity should be removed by definite recital such as "center line of ____ Street ____ feet wide" or similar form. *The same rules apply if the street is narrowed by vacation.*

3 "transit line" may apply to any surveyed reference line which is not a true line, and should be particularly defined if the term is used.

4 without qualification, must be assumed as a "center"; always specify.

Central angle; intersection angle; delta; through an angle

Synonymous, and equally good from an engineering standpoint; however, "central angle" seems preferable in most cases, as being most easily understood; "through an angle" more particularly defines the angular measurement of an arc.

Circle; arc; curve; center of each (Fig. 5)

A circle is the area included within the circumference described as the line produced by a complete revolution of the extremity of a radius of said circle about a fixed center. Circumference is sometimes considered as, and stated (erroneously) as "circle."

The center of the circle is the fixed extremity of the radius about which rotation occurs.

An arc of a circle is a segment of the circumference of the circle.

The center of the arc is the mid point of the arc between the extremities of the segment.

A curve is generally understood to be synonymous with an arc, although it may be applied to a group of continuous arcs of a boundary.

The center of a curve is the mid point of a single arc. The term is sometimes erroneously applied to the center of the circle. Applied to a group of successive curves, it would be the mid point between the extremities of the total group: Such a reference should be further qualified for safety.

1 convey; 2 describe 900

1 means a definite transfer of the title, and is applicable correctly to the area transferred, which may include land not directly described or only land remaining after exceptions are had; a description reciting "the NE corner of the land conveyed to ________" may call for a corner created by an exception or addition and not the corner intended.

1 is frequently used as synonymous with 2 and is a hazard.

2 refers to the particular thing stated or recited in the description to which reference is made, and may or may not be directly connected with the "conveyance"; *e.g.*, the strip "described" for a road may not be the strip "conveyed," if the grantor owned no part or only a small part of the strip.

A description calling for "(such and such) a part of the following described land: . . ," following with description of the larger parcel, including exceptions, must be construed as covering the part called for, of the *net* area produced by taking the exceptions out of the larger parcel; if it is desired to take the exceptions from the "part," that fact must be stated clearly. This is the so-called "double exception" form, and is very dangerous unless adequately handled; there is generally a safe way out by use of another type form.

Course

This includes the dimensions of the line, both bearing and distance. If segregation is desired, use "bearing and distance" *not* "course and distance." The term may in some cases apply to curves.

It is common usage to consider "course" and "bearing" as synonymous (from the mariner's language) and is passable.

1 Described as follows or described as; 2 described as a whole as follows; 3 more particularly described as follows

1 Normally used as a conjunctive between caption and description.

2 Preferably used when the one description is to include the several different portions of the caption, although 1 may be used if the meaning is clear.

3 Often used as a substitute for 1 but more specifically applied as the preface of a second description similar to and intended to cover the same land as the first. It is also used as a preface to an accurate description supplementing a general statement such as "The S½ of

900 the E½ of Sec. . . . more particularly described as follows (and follow with a metes and bounds). . . ."

A distance of

Used to separate different kinds of numbers and to prevent errors in distances, such as: "to the corner of said Lot 5 a distance of 3.27 feet," instead of "to the corner of said Lot 5, 3.27 feet." Not necessary except in such cases, or in cases where the distance is separated by several words, or a phrase, from the line whose length is given.

Due (North, East, etc.)

Because of ambiguity and uncertainty in meaning and use, DO NOT USE this word in descriptions to indicate bearing direction.

The word is often used as a modifier of the cardinal directions. It is sometimes given as a synonym of "true," "astronomical," or "magnetic" North (or East, etc.); more probably it is interpretable as a qualifying term to distinguish the bearing (North) from the general direction sense (North, or North along a line).

There is no definition showing "due" as meaning astronomical or similar connotation, but custom, use, and context may on occasion indicate the intent.

"Due" is not used in bearings other than the cardinal.

If interpreted as an astronomical call, "due" is "true" only on a North or South line. "Due" East or West will define a parallel of latitude, a curve, not contemplated in planimetric surveys. An "East" line in the planimetric system is a *straight* line at right angles to the reference (base of bearings) meridian of the survey in question.

If used in a planimetric survey description, unless qualified, "due North" defines a line *parallel* with that single reference meridian in the description accepted or assumed as "North."

Except; reserving: subject to

"except," means exclusion from the area, of the part excepted; usually applied to a fee title.

"reserving," is a taking back of part of something granted; usually applied to easements, or imposed incumbrances of various sorts.

"subject to," refers to a reservation already existing, as distinguished from "reserving" which creates in the instant case.

A described area carries "reserving" if the easement is therein created, or "subject to" if the easement already exists and is mentioned as an incumbrance. 900

"except" or "subject" as applied to the effect of an easement, incumbrance or fee parcel on described land implies precedent existence.

"reserving" applied to the effect of an easement or incumbrance on described land implies a creation of such easement or incumbrance in the instant description.

Except the interest of . . .

Used in excluding a right of an owner other than the grantor or grantee in an instrument; particularly applied when the status of the area affected by the exception is doubtful as to its being a fee or an easement. The exception of the "interest" will apply to the tenement actually affected.

1 Extending; 2 prolongation; 3 continuation

1 to be applied to a line being created, a part of which is a portion of the boundary of description;

e.g., ". . . to a point in a straight line extending from the NW corner of said Lot, Ely to a point in the E line. . . ."

2 to be applied to the continuation of a line already existent.

3 and 2 often used interchangeably but more specifically a straight line is "prolonged" and a curve is "continued." (See also "prolongation etc." post.)

1 first above; 2 last above

1 applies to the first statement in the description of the item mentioned in the instant phrase.

2 applies to the last immediately preceding statement in the description of the item mentioned in the instant phrase;

e.g., in the description ". . . at a *point* in the North line of Lot A; thence Easterly to a point in the West line of 6th St., distant 100 feet Nly from the SE corner of Lot 6; thence SEly to the SW *corner* of land of A, from which the first above mentioned *point* bears N 40° W 327 feet; thence from said last mentioned *corner* N 89° E. . . ."

The italicized *points* are the same and the italicized *corners* are the same.

900 ***Front and Rear***

Used to define the dimensions at each end of a lot or part, implying the "front" as the length on the street or easement line and the "rear" as the length on the portion of the boundary opposite the "front."

Obviously uncertain if applied to a corner lot with adjacent street frontages; also indeterminate if applied to a lot having double street frontage, *i.e.*, frontage on each end of the lot.

Not good form. Much better and safer to recite the directional position of the boundaries dimensioned, such as "the Wly 50 feet measured along the north and south lines of said Lot."

1 hereinbefore; 2 hereinafter

Refers to the stated item which precedes (1) or succeeds (2) the matter recited in the instant phrase.

1 lot; 2 piece; 3 parcel

1 may refer to a particular area designated by letter or number on a record map.

3 generally used to designate a particular numbered or lettered area on a Record of Survey Map, Parcel Map or a non-record, or quasi-record map to distinguish it from a true "lot" (1).

2 nearly always used with 1 and/or 3 and has no value as a designator, for either 1 or 3 is sufficient; redundant and superfluous.

1 and 3 often used as a general designation or caption for an area not specifically nameable, such as "a parcel of land (or—a lot—) in the City of ______________."

1 lying; 2 being; 3 situate; 4 included

1, 2 and 3 used together in the old form, provide an unnecessary redundancy, for they are in this combination synonymous. Preferable to state "lying within," "being a portion," or "situate in" or some similar form.

2 is the preferable word of locale, unless 4 is necessary for distinction.

4 used especially when only a portion of an area described or referred to is to be considered.

Mid point; middle point; point midway; bisection (point)

Synonymous terms as applied to a line having a definite dimension, or termini; even if the line is not straight, designating a point on the line equidistant between the termini of the line.

Bisection applied synonymously with mid point is not recommended as a line may correctly be bisected into unequal parts.

1 More or less; 2 approximate; 3 about; 4 plus or minus

1 indicates uncertainty as to accuracy of dimension but in small degree; may be used to distinguish an uncertain or secondary call from a definite one. Should *not* be used with a bearing as "N 60° E more or less." Also used to differentiate between the exact and the secondary tie in a double call.

2 indicates ignorance of exact dimension, but a knowledge within reasonable limits. May be used with bearings, as "approximately N 60° E."

3 indicates very general information and should be used only for broad identification. May be used with bearings, as "about N 60° E."

4 sometimes used as substitute for 1; not recommended.

When 1, 2 or 3 are used without tie or reference, they are meaningless and the dimension stated in connection with such phrase is to be taken as exact: *e.g.*, "East 230 ft. more or less" means exactly 230 feet, as no tie is given.

1 North; 2 Northerly (and other directions with or without "-ly")

1 may be true bearing or sense of direction depending on its use with relation to rest of subject matter.

2 usually implies sense of direction but not specific bearing; however, construed legally, if the contrary does not appear, means *true* (meridian) North.

2 should *never* be used as a part of a bearing (Nly 70° Wly) (correct: N 70° W).

2 preferably used in descriptions to designate sense of direction, especially if certain courses in descriptions carry specific bearings of "North," "East" etc., or if the directional sense is considerably different from North, etc. (such as N 30° E).

Parallel

Two straight lines, or their prolongations, at equal distance apart (*i.e.*, having the same bearing on a map or in a description) regardless of their condition of adjacency or position, are parallel. (Fig. 4.)

900 Two lines, each composed of two or more segments of differing direction are, as a unit, parallel, if each segment of one is at the distance measured at right angles, from the corresponding segment of the other as every other segment is to its fellow; it follows that the corresponding angle points of each line respectively, lie on the common bisection line of the angles of which those points are the vertices. (Fig. 8)

Curves are parallel if they are concentric; generally applied to arcs lying approximately adjacent and in the same general direction from the common center of the circles of the arcs. (Fig. 6)

Lines may be parallel as regards their respective segments, but not parallel as aggregate units, if the above condition of equal distance between all corresponding segments is not maintained. (Fig. 8)

(See also section on parallelism ante)

Point; to a point; at a point; point of intersection

The word is too frequently used without qualification and as such is without merit. The end of any course, or the intersection of any two lines *must be a point* so there is no need to state it. If the point has special features, or references, or is to be referred to later, then it should be recited.

1 Prolongation; 2 continuation; 3 projection; 4 extension

With reference to a line, 1 is used to indicate the continuity in the same direction (bearing) of the recited course which is to be lengthened; 2 is used if the line is a curve, and the continuation (prolongation) is the extension of the arc in the direction specified, with the *same radius* and *concentric*. A "prolongation" of a curve is strictly an extension in the direction of the tangent of the curve at the point of extension and the term should not be used in referring to curve continuation. 3 and 4 indicate general direction of continuity and are not necessarily as rigid as 1 and 2, although sometimes used synonymously.

The prolongation of a line or boundary, composed of several parts of different directional values, is the prolongation or continuation of the course or curve nearest to the recited intersection or monument.

1 Recited; 2 noted; 3 described

1 is usually a verbatim quote of a statement in a deed, map or other reference; *e.g.*, "N 60° W, recited in said deed as N 60° 10′ W."

2 used as reference to a matter with or without confirmation as to correctness; as "to a stake noted as the NW corner of Lot 6 on map of Tract . . ." 900

3 is rather inclusive in its meaning and properly intended to refer not only to the particular item stated but also to the related matters connected with it and defining it; *e.g.*, "to the NW corner of the land described in deed recorded . . ." implies the correct position of the corner as developed from all the matters in the deed which may establish it.

Right angles; perpendicular; normal; radial (as applied to curves and tangents thereof)

All expressions denote the relation between two lines having an angle of 90° between them.

A straight line may be stated as: "at right angles to," or "perpendicular to," or, "normal to," another straight line; "right angles," is the preferred term.

A line "normal" to a curve is the *prolongation* of a radius of a curve (*not* the radius itself) beyond the circumference of the curve; the use of "along a radial line (or, along the prolongation of a radial line)," is more accurate, and to be preferred to the "normal to said curve."

The radial line (or radius) of a curve is at right angles, or perpendicular, to a line tangent to the curve at the point of tangency.

Right, left

When applied to angles or curves indicating change of direction from last preceding course or curve, is old form and not satisfactory because of need of interpretation. The use of such terms applied to the banks of a meandering stream is satisfactory by reason of definition of the terms, but when applied to particular segments of a traverse, may be uncertain, since a re-entrant angle or cusp may exist in the boundary and the course can properly be described by "thence to the right," whether extended ahead or in the opposite direction; the best method of deflection recital is by direction of bearing such as Northerly, Southeasterly, etc. (Fig. 3)

Running

Superfluous; does not add to meaning.

900 ***1 Said; 2 aforesaid***

1 refers back to the next immediately preceding item of the same name unless qualified, but not to a succeeding item. Any qualification must be specific to prevent uncertainty.

2 nearly synonymous with 1, but broader, possibly covering more than one item; preferably omitted in favor of 1.

Tangent

A straight line having a curve contiguous at one or both extremities to which the line is tangent; specifically qualified when used in descriptions, such as

1. "Tangent portion of the North line of Lot ____"

   applied to the straight line part of a lot having a curved corner or boundary on a portion of the "North" (northerly) line.

2. Tangent *to* said curve

   applied to a course of boundary which is tangent to the recited arc of a curve.

3. Tangent *of* a curve

   generally refers to the extension of a tangent to a curve from one extremity to its intersection with the corresponding extension of the tangent from the other extremity of the same curve. Frequently called the "semi tangent."

4. Tangent

   In railroad practice the straight line portion of the alignment which begins or ends in a curve tangent thereto.

5. Tangent curves

   Reverse or compound curves. They have a common tangent (a line perpendicular to the common radius) at the point of reverse or compound.

6. Non-tangent curves, courses, or lines

   Not possessing the quality of tangency with or to the next or adjoining curve, course, or line.

   Applied only as between straight lines and curves, or between curves.

1 That portion; 2 a portion; 3 those portions; 4 any portion

1 and 2, interchangeable, but prefer 1 as more specific.

3 to be used when the description covers more than one particularly designated piece of property (those portions of Lots 6 and 7), but not (those portions of Lot 6), unless the description consists of two or more parts, each of which describes a separate part of Lot 6. 900

3 also used as caption for separately designated description such as Parcel 1, Parcel 2, etc.

4 used when there is a doubt or question as to whether there is any part of the described land affected by or affecting the portion defined.

1 Thereof; 2 therefrom; 3 therein; 4 thereon

Each refers back to a previous item in the description as an identification.

1 and 2 used interchangeably, also together in the same phrase; *e.g.*, "excepting therefrom that portion thereof . . ." (superfluous and not recommended). 1 also used to particularize an exception as applying to a certain part of the description; *e.g.*, "Lot 6 except the E 50 feet thereof, and all Lot 7. . . ."

3 and 4 generally interchangeable, referring to position in or on a line; frequently used too freely and incorrectly from a grammatical standpoint; for example:

1. "Thence along the North line of said Lot to a point distant Easterly 200 feet from the Northwest corner *thereof*."

   Thereof refers correctly to the *line*, not the lot and there is no corner to a line. The word "thereof" should be replaced by "of said Lot."

2. "To the point of intersection of the South line of A Street with the East line of 14th Street distant *thereon* . . ."

   Thereon refers correctly to the point of intersection and a distance cannot be measured along a dimensionless point. The word "thereon" is unnecessary and should be omitted; "distant along said line" should be used.

With; to; from; at; through; from which; at which

A line is parallel WITH (not "TO") another.

A line is at right angles TO (or "FROM") (not "WITH") another. "With" means association; "to" and "from" are directional.

The sense of direction of the radial line of a curve is correctly stated from the center of the circle toward the circumference. This

900 direction should always be used when giving the bearing of a radial line "to," "through" or "at" a point in the curve; it should be reversed when reciting the bearing "from" the point in curve.

At point "A," *from which bears* point "B," carries a bearing sense from A to B.

At point "A," *which bears from* point "B" carries a bearing sense from B to A.

"At which" is used in similar sense with "from which."

In general, any abbreviation or character which is in common use, and definitely identified with a particular word or phrase, is legally usable in a description, as the use of such abbreviation or character will not tend to destroy or render uncertain the meaning of the description. However, safety demands the full words in cases of uncertainty or infrequent use of the abbreviation, or use of an abbreviation common to more than one term (such as L C for long chord or length of curve). Maps may carry abbreviations more readily than descriptions because of easier interpretation.

The figures of a number are usable instead of the words, although if both are given and one differs from the other, the words will prevail unless the contrary can be proved.

Bearings may be written with the initial letters of the direction and the figures with the engineering symbols for the quantity of degrees, minutes, and seconds (N 80° 10′ 37″ E).

All bearings except the cardinal ones (N, S, E, W) should show degrees, minutes and seconds, substituting zeros if no finite quantity is given (N 0° 00′ 06″ E), using one zero for degrees and two for minutes or seconds, with one preceding the minutes or seconds as the case may be if less than two significant figures; it is unnecessary, however, to carry zeros in the minutes, or seconds, if there is no subsequent quantity (10°; 8° 05′). It is dangerous, even if correct, to omit the designation of zero quantities. "26 seconds" may be so written, safely, but not 26″; it might be construed 26 inches, or if shown N 26″ E be blurred or otherwise changed to N 26′ E or N 26 E and possibly considered as 26°.

The cardinal directions are preferably written in full (North, West, etc.).

The foot (′) mark and inch (″) mark are usable but, because of the similarity of minutes (′) and seconds (″) designation, preferably use the word "foot" or the abbreviation "ft." and "inches" or "in."

Chains and varas may be used as "ch." and "v." although "vara" is preferable.

Center line may be noted as "c.l." or "L" (uncommon except on curves on maps).

"A" may be used, but "Ac." or "Acres" is preferable.

The intersection angle or "delta" of a curve may be shown as "Δ" on maps, but prefer the words in a description, such as "central angle," "intersection angle" or "through an angle of."

Curve dimensions on maps may be shown as L (arc length), T (semi tangent) R (radius), but should be written out in descriptions.

1000 Directions may be abbreviated by their initial letters as "N" (north), "Ely" (easterly) "SW" (Southwest), "NWly" (northwesterly).

On maps, P. L. (property line), S & T (spike & Tin), 2″ p (2 inch pipe), conc. mon. (concrete monument) etc., may be used, but descriptions should preferably use the full words.

All decimal dimensions, less than 1.00 (one) should be shown with 0 in the unit place (0.27), to make certain the decimal point is recognized.

On maps with dimensions in feet and inches, safe to use characters (27′ 10″) except where close by bearings may cause confusion.

With dimensions in units and hundredths, the character may be omitted if the scale of the map is given in the stated unit. (100.25)

If more than one type of unit is used on the map, the scale unit needs no character mark, but all others should be stated (scale 1″–100′; dimension for a line is *329.45*, to represent feet, or *5.16 ch.*, to represent chains, etc.).

It is dangerous and very unsatisfactory to represent decimals by an underline and the character, as 504′; the ′ mark may be mistaken for a "one" or the underline may be so drafted that the whole distance may become 5.04 or 504.

It is unsatisfactory to write decimal figures in fractions, especially if the decimal point is also included, such as 124.6/100 ft; it may mean 124.06 or 124.60 on the assumption that the "0" after the 6 was inadvertently omitted.

Any quantities either on map or in description are preferably stated in units and decimals rather than units and fractions.

Unless quite apparent, it is well to use the word together with the character in Block 1 (one) or Block I (eye); or in referring to a record using the Roman numerals, such as "Lot III (3) of the partition of Block H of the Rancho Los Palos Verdes." Quotation marks for distinction of special letters or numbers may be used, as Block "I," Lot "III" etc. This precaution is especially necessary in certain forms of typewriter letters.

"City and County" of Los Angeles does not legally exist; use "City of and County of Los Angeles" or preferably "City of Los Angeles, County of Los Angeles." However, it is correct to say "City and County of San Francisco," or "City and County of Denver," because of the Legislative Act creating such a unit.

Punctuation marks and typographical arrangement

No description should be dependent on punctuation for its interpretation. Punctuation should be used as a matter of course, and clarity, and separation of items, but not for the purpose of segregation or aggregation, when the *omission* of such marks could definitely change the meaning. The perfect description is one which can be positively identified in meaning even though written *without any* punctuation, the variances and phrasing being so selected that they separate the parts by their own structural differences. 1000

Free use of separative and conjunctive words or phrases is recommended, such as "the," "and," "also," "said," "last mentioned," "firstly described," "thereof," and so forth, but not to the extent of redundancy.

A certain amount of paragraphing and separation of statements is advisable for ready reading, but the use of such methods for specific variants in meaning is dangerous; a recopy with another form of setup develops a hazard. Any description dependent on typographical arrangement for its interpretation is faulty and liable to uncertain construction.

EXAMPLE: (1) Parcel 1. The East 60 ft. of Lot 6, Tract M
Parcel 2. That portion of Lot 8 adjoining said Lot
(original) 6 on the north.
Except the East 20 ft.

(2) Parcel 1. The east 60 ft. of Lot 6 Tract M.
Parcel 2. That portion of Lot 8 adjoining said Lot 6
(as copied) on the north. Except the East 20 ft.

The typography of (1) normally implies the exception to apply to each parcel though it might not have been so intended, while in (2) it would apply only to Parcel 2. Each description is intended to cover the same land; which interpretation is correct?

In writing or typing do NOT split bearings or angles on successive lines such as:

N 27° – 46′ or S 62° – 25′ or 46° –
30″ W E 27′ 24″

It is incorrect to the same extent as splitting distance such as:

20. or words Str
72 ft. (for 20.72 ft.) eet (for street)

Phrasing arrangement

Care must be taken in arrangement of statements of reference or modification of prior statements. The following example will illustrate:

1000

1.	Lot 5 Blk 2 except the E 5 ft	correct
2.	" " " " " " E 5 ft of said lot	correct
3.	" " " " " " E 5 ft *thereof*	might apply to Blk 2 and not to Lot 5
4.	Lot 5 except the E 5 ft of Block 2 —	incorrect; can apply to Blk 2 and not to Lot 5
5.	" " " " " " " in Blk 2 —	correct
6.	" " " " " " " thereof of Blk 2	passable

1100 CALCULATIONS AND SHORT METHODS

No satisfactory or successful work can be done with description or survey of land and titles without a reasonable knowledge of the mathematics connected therewith, and the ability to make fairly accurate plats. The amount of calculation and extent of mapping depends on the complexity of the problem.

A ready use of short methods, and quick approximations will assist greatly in many situations.

A file of formulae, approximations, tables, etc. is a valuable adjunct.

For rough field check without instruments

1. Angle from eye subtended by edges of fist held horizontally with arm fully extended forward is approximately 8°.
2. Angle from eye subtended by tip of extended thumbs turned inward on arms extended squarely forward from body approximately 10°. Fig. 2a
3. Angle from eye subtended through tips of middle fingers on arms and hands extended squarely forward from body approximately 30°. Fig 2a
4. 90° angles to a line A lies on line of sight over hands together directly in front of body when feet placed squarely on the line A extending sideways from body.
5. *Grade* Sight over tip of thumb extended vertically, with arm extended forward horizontally; note intersection point on ground and pace to point. Grade percentage is approximately one divided by half the number of paces. Fig. 1a

Angle relation to distance (approximate)

OFFSETS

0° – 01′ gives spread between line of 0.03 per 100
0° – 35′ " " " " " 1 per 100
5° – 45′ " " " " " 1/10 of distance observed
30° " " " " " 1/2 " " "

LENGTH

0° – 35′ gives horizontal distance less than slope of 0.005 per 100
2° – 30′ " " " " " " " 0.1 per 100
5° – 45′ " " " " " " " 0.5 per 100
8° – 0′ " " " " " " " 1 per 100
30° " " " " " " " 15 per 100

CONVERSIONS

Chains equal feet × 1.5151
Feet equal chains × 66
Acres equal square feet × 0.000023—result is too big by 0.2 per 100
Square feet equal acres × 43560

1101 Acres

1 acre 43560 square feet covers a square of 208.71 feet on a side
0.1 " 4356 " " " " " " 66 " " " "
0.01 " 435.6 " " " " " " 21 " " " "
0.001 " 43.5 " " " " " " 6.6 " " " "

For approximate area of a quadrilateral see Fig. 3a

Fig 1a

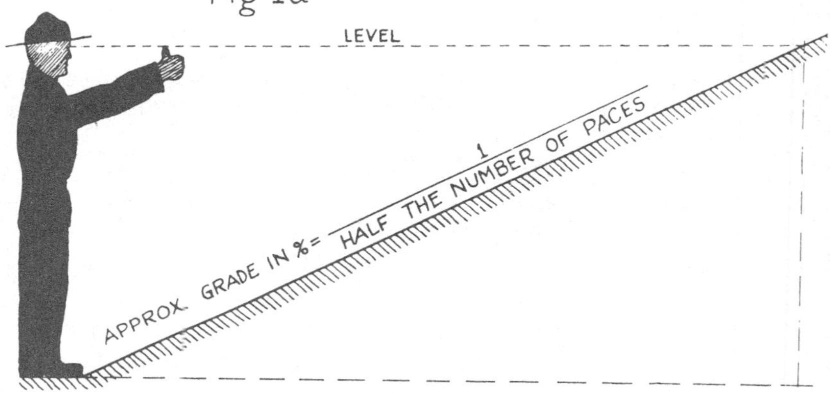

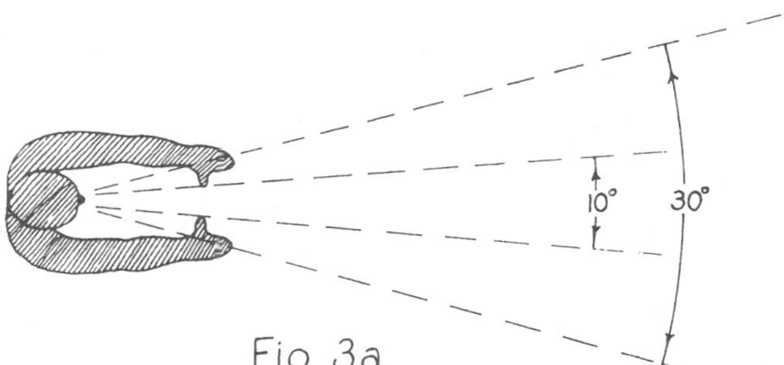

Fig 2a

Fig 3a

Approximate quadrilateral area

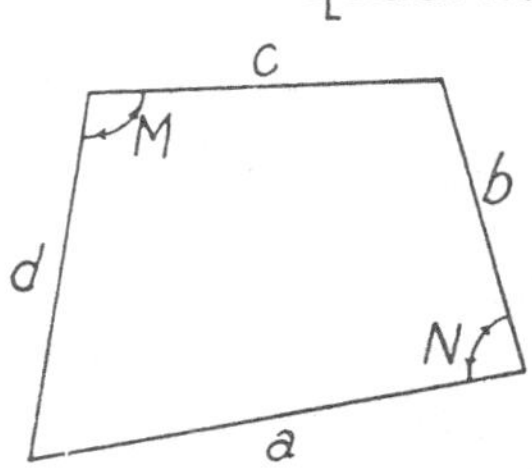

Diagonally opposite angles M & N each not more than 4° from 90°–

Area = (ab + cd) x 0.0000115, in Acres
The result will be too great by approximately $\frac{3}{1000}$.

Legal Controls of Descriptions, Maps and Surveys are founded in law governing the specific activities of surveying, subdividing, mapping, rules of construction, etc. The following are references to some of the subject matter including the new references to the Government Code for the Map Act of 1974.

	Code Sections*
Amending Map	GC 66469
California Coordinates	PRC 8801
Cemetery Exemption	GC 66412 (c)
Certificates	GC 66435, 66445 (a), 66449, 66450
Certificate of Compliance	GC 66499.35, .36 & .37
Certificate of Correction	GC 66469
Condominiums	GC 66426, 66427
Condominiums	CC 787 & 1350
Corner Record	BPC 8773
Dedication Abandonment	GC 66477.2 (c) & (d)
Dedication Offer	GC 66439, 66477 (b) & (e), 66477.2 (a) & (b)
Dedication Termination	GC 66477.2 (e)
Earthquake Act	CCP 751.50
Exclusions	GC 66499.21-.29
Final Maps	GC 66426, 66433-66443, 66456-66462
Land Surveyors Act	BPC 8700
Monuments	GC 66495-66498
Notice of Violation	GC 66499.36
Official Maps	BPC 11650
Official Maps	PC 3658a (prior to 1943)
Parcel Maps	GC 66426, 66428, 66445(a)-(g), 66447, 66448, 66449(a) & (b), 66450, 66463, 66463.5(a), 66464(a) & (b), 66466(a), (b) & (c), 66468, 66477.1 & .2 & .3
Record of Survey Maps	BPC 8762
Reversion to Ac. Maps	GC 66499.11-.20½
Rules of Construction	CCP 2077
Subdivision Maps	GC 66424, 66426, 66445, 66464(a) & (b)
Subdivision Maps—Non	GC 66412
Tentative Maps	GC 66452, 66452.1-.7, 66453, 66454, 66455.5 & .6, 66456, 66474, 66464.60 & .64

*BPC: Business & Professions Code
PRC: Public Resources Code
CC: Civil Code
PC: Political Code
CCP: Code of Civil Procedure
GC: Government Code

Refer to Statutes of California for the various Map Laws passed from time to time. The recordation of subdivision maps originally operated under the provisions of the Business & Professions Code by the Act of 1937 and amendments thereto, including the "Local Ordinances" permitted and defined in the Act. Effective March 1, 1975, the above was recodified into the Government Code by Chapter 1536 of 1974 Statutes and by Chapter 24 of 1975 Statutes effective April 4, 1975. Further amendments were enacted in 1975 through 1980. Read the Law and Ordinances, and obtain information from the Planning Commissions as to procedure thereunder.

The Local Ordinances permitted under the Subdivision Law include certain regulations created under authority of Cities or Counties beyond those required by the Map Law.

The attempt of the Map Law is to prohibit subdivision and recordation of maps unless certain regulations as to use, as well as plan are followed. However, the Law specifies certain exceptions, in definition of "subdivision," and in the use of certain maps already on file or of record, or filed in conformity with other Acts providing for filing of maps.

Recordation of a map does not of itself change the status of fee title of the land covered; certain easement titles may be created by dedication. Any change in fee title must be made by proper conveyance.

Good record conveyance can be made by reference to a map, recorded or filed; Tract Map, Parcel Map or Record of Survey Map. A conveyance may be made on the title sheet of a record map by the proper use of proper words.

Do not show private easements (*i.e.*, easements created between private parties) on record maps. However, any easements required by governmental bodies to be offered for dedication must be shown.

Memoranda or other information which will render more certain any locations shown on a map should be placed thereon, especially data indicating the manner of establishment of a record boundary. Correct offsets and pluses to found monuments near, but not on, a line, are also essential.

Although showing on a record map is permanent, a new subdivision map covering the same area may supplant the old, or an Amended Map may correct certain errors, and proper conveyance may be had by reference to the new map. This is an excellent method for correcting errors on the previous map, since the newer map will act as a proper substitute, independent of any discrepancies in the former. Filing of affidavit of error, with correction, or filing of a Record of Survey showing correction of error is informational and may produce estoppel; it is good practice to

make reference in conveyance to such forms of correction, such as 1200
"Lot _____ etc., the dimensions thereof being modified by affidavit (or Record of Survey) recorded (or filed)... etc."

It is competent, and occasionally preferable to write descriptions referring to the older map, and data thereon, regardless of resubdivision; in such case, a modifying phrase should be added such as "Said land is also shown as portions of (*naming the new tract with its reference*)" or "said land with other land has been resubdivided and is shown on tract _____ recorded in (*map reference*)."

(Torrens) Land Registration Law

This was essentially a quiet title proceeding, and when properly conducted, with a correct description, defined the boundaries of the land whether conformable with previous record or not.

Specific information and description of registered land could be obtained from the Registrar of Titles. The form and content of descriptions for registered land had to be acceptable to the Registrar before filing. Instruments were drawn in duplicate.

Any instrument filed in the Registrar's office could, if desired, be recorded also, in the Recorder's office; sometimes this was advisable. If the description covered both registered and unregistered land, there had to be a definite segregation of the areas, and the instrument had to be both filed and recorded, to cover both types of land.

General description and location procedure was the same as for unregistered land.

The law originally made no provision for freeing land from registration when once placed under the system, but legislation in 1952 provided a means of reconversion of title to usual recordation procedure, and in 1955 returned all registered land to the recording system, including transfer of all files and registration records.

Land Surveyors' License Law

Allows a L.S. to perform any and all types of surveying and location, make maps, render reports and set monuments, but *prohibits design* of structures or fixed works.

A Registered Civil Engineer, if qualified, may also practice land surveying.

Requires the filing of a Record of Survey map under certain conditions, particularly where "material discrepancy" with the record

1200 occurs, or where "alternate positions" of a line or point are logically possible. Advisable to file map where perpetuation of monuments is desirable, such as those in impermanent position, or of an age where destruction is imminent; such a filing is an advantage to the surveyor in many ways. The map filed is what its name implies – a record of a survey – and should be *complete* as to information of such survey, either by direct showing on the map or by notes and explanations thereon; monuments may or may not be set, those shown as "found" being sufficient in many cases to evidence the purpose of the map.

A Record of Survey map does not change title except when used as: Agreement Map; formerly in lieu of a subdivision; or by Court order.

Wherever durable monuments are set to mark or reference a point on a property or land line, they *must* be visibly marked or tagged with the L.S. (or R.C.E.) number.

Whenever any formal or final document prepared by an L.S. or R.C.E. is delivered as such or filed as a public record, it *must be signed* and bear the number and seal of the L.S. or R.C.E.; this applies to sketches, maps, descriptions, reports, etc.

Official Maps

Filed by authority of Board of Supervisors under Section 3658a of the Political Code (prior to 1943) or by authority of Board of Supervisors or other governmental authority as per Sections 11650 to 11658 of Business and Professions Code (after 1943).

Official Maps may be compiled from record covering only existing parcels of land already surveyed or conveyed. There is no authority in the Code to allow the Official Map to be used to create a subdivision of new parcels; questionable whether permissible to alter record dimensions of parcels by check survey although some Official Maps purport to show information obtained from survey in comparison with or contradiction to previous records.

Pursuant to a court decree in accordance with the Earthquake Act, an Official Map shall be filed as a substitute for plat maps previously filed.

Other Map Laws

Provide for filing or recording maps, plans, surveys, for special purposes. Each law establishes the procedure peculiar to its needs. Refer to the particular law for information governing the case.

The following index covers references to laws pertinent to Descriptions, Maps and Surveys.

	Chapter Sections
Amendment of Plats, Surveys & Maps	278.473 et seq
Certificates	117.027, 278.373 et seq & 625.330
Condominiums	117.010 et seq
Corner Perpetuation and Record	329.010 et seq
Correction Maps	270.010 et seq 278.473 et seq
Dedications	278.390
Final Maps:	
Condominiums	117.020, 117.025
Corrections	278.475
Record of Surveys	625.350
Subdivisions	278.360, 278.466
Land Surveyors' Act	625.250 et seq
Monuments	278.400, 278.371, 625.380
Nevada Coordinate System	327.010 et seq
Parcel Map	278.461 et seq
Records of Survey	625.340
R.S. Maps	625.350
Reversion to Acreage	278.490
School Sites	278.346
Subdivisions	278.320 et seq
Survey & Map by C.E. or L.S.	278.360, .371 & .463
Tentative Maps	278.330 & .464
Title in Streets	278.390
Vacation of Streets or Easements	278.480 et seq

Subdivision Map Law

Refer to the Revised Statutes of Nevada for the various Map Laws passed from time to time. The recordation of subdivision maps operates under the provisions recited in Chapter 278 of the Nevada Revised Statutes. Read the Law and Ordinances, and obtain information from the Planning Commissions as to procedure thereunder.

If the land is situated in any city or town, or within 1 mile of same, the map must be approved by the legislative authority of the city or town in which the land is situated, and in the absence of such legislative authority, by the legislative authority of the county in which the city or town is located. If the land is situated more than 1 mile from the limits of any city or town, the map shall be approved by

1201 the board of county commissions of the county in which the land is situated.

The attempt of the Map Law is to prohibit subdivision and recordation of maps unless certain regulations as to use, as well as plan are followed.

Local Ordinances may be enacted to govern subdivision improvements and related subjects. Recordation of a map does not of itself change the status of fee title of the land covered; certain easement titles may be created by dedication. Any change in fee title must be made by proper conveyance.

Good record conveyance can be made by reference to a map, recorded or filed.

Do not show private easements (i.e., easements created between private parties) on record maps. However, any easements required by governmental bodies to be offered for dedication must be shown.

Memoranda or other information which will render more certain any locations shown on a map should be placed thereon, especially data indicating the manner of establishment of a record boundary. Correct offsets and pluses to found monuments near, but not on, a line, are also essential.

Registered Land Surveyors' License Law

Allows an R.L.S. to perform any and all types of surveying and location, make maps, render reports and set monuments.

Requires the filing of a Record of Survey map where certain conditions, particularly "material discrepancy" with the record occurs, and "alternate positions" of a line or point are logically possible. Advisable to file map where perpetuation of monuments is desirable, such as those in impermanent position, or of an age where destruction is imminent; such a filing is an advantage to the surveyor in many ways. The map filed is what its name implies – a record of a survey – and should be complete as to information of such survey, either by direct showing on the map or by notes and explanations thereon.

Monuments of sufficient number and durability are to be set and tagged.

Any written or documentary evidence, or maps, of a survey or division of land, or of matters pertaining to land location must be presented to, filed or recorded with the proper legal governmental authority as a permanent record in order to provide for the proper use of such evidence as a reference in other documents.

The purpose of recording or filing documents in public offices is two-fold: to perpetuate the information contained in the documents, and to give notice to the world at large, enabling interested persons or the courts to operate and conform with the contents and provisions of the transactions set forth in such documents.

Documentary matter will include deeds, mortgages, affidavits, notices of all kinds, tax matters, court files of pending or completed cases, ordinances and resolutions of Governmental Bodies, maps, plats, diagrams of any nature, or any other material affecting land holdings, transfers, incumbrances, or titles.

The place and manner of recording a document is determined by the laws of the particular District or State in which the property covered by the documents, or affected thereby, may be located. Federal laws specify the manner of recording matters affecting federal activities.

Surveyors or other persons writing descriptions or making surveys and locations therefrom, are bound to familiarize themselves with the various types and methods of recording and filing which affect the data applicable to the area under consideration.

A true record is one properly presented and filed or recorded with the designated authority as provided by the laws governing the establishment of "Records."

EXAMPLE: A map of a subdivision recorded in the Recorder's office is a true record; a map of a road alignment filed in the County Surveyor's office is a quasi-record.

Any reference in a document or map to a record instrument, reads into such document or map all the matters applying thereto which the record instrument may contain.

Although a document may be valid and effective per se, the recordation must be made conformable with the prescribed law in order to provide a legal record.

Certain documents and maps may be accepted and filed in public offices, or may originate therein and carry a file designation for identification, which are not classifiable as proper legal records, but may

1300 be termed quasi records; these are available to public inspection but are not good as proper references in description.

Many proper records or quasi records are copied wholly or in part by offices or persons for personal reference, and given identification pertinent to such particular office. Reference in description to such identification is unacceptable since the material is susceptible to being changed, and no assurance can be had of the accuracy of the copy.

Documents on file in non-public offices, and not open to general inspection by the public, are not acceptable as reference material in description writing; special circumstances, however, may warrant a reference to a map prepared by some public utility office (railroad, telephone, electric power, etc.), if no other factual material is available.

A proper description must contain correct and proper references to the various records; a surveyor must make use of such records for proper interpretation in location.

Although a reference improperly recited as "recorded" instead of "filed" (or the reverse), may under some circumstances be accepted as passable for identification; such improper recital indicates negligence or ignorance in the drafting of the description.

Complete and correct reference is necessary to maintain accuracy and certainty of interpretation of a description. Reference will include: Name and nature of the referred document (deed, map, field notes, court partition, etc.); place of record or filing (State, County, City, District, etc.); the public office wherein the document is presented and accepted (County Recorder, City Engineer, County Surveyor, Board of Supervisors, County Clerk, Registrar of Deeds, State Highway, County Flood Control, etc.); the nature of disposition (recorded, filed, attached ((to some other document)), etc.); the book and page of record, file number, map name or number, court case file number; when deemed necessary, the date of the instrument or the date of record; occasionally, though not always necessary, the name or names of parties to the document (grantor, grantee, transferee, surveyor, official or body approving map, etc.).

When incorporating any record reference in a document or map, be certain that the correct terminology for such reference as to place and kind of record, is stated.

DESCRIPTION VARIANCE

1400

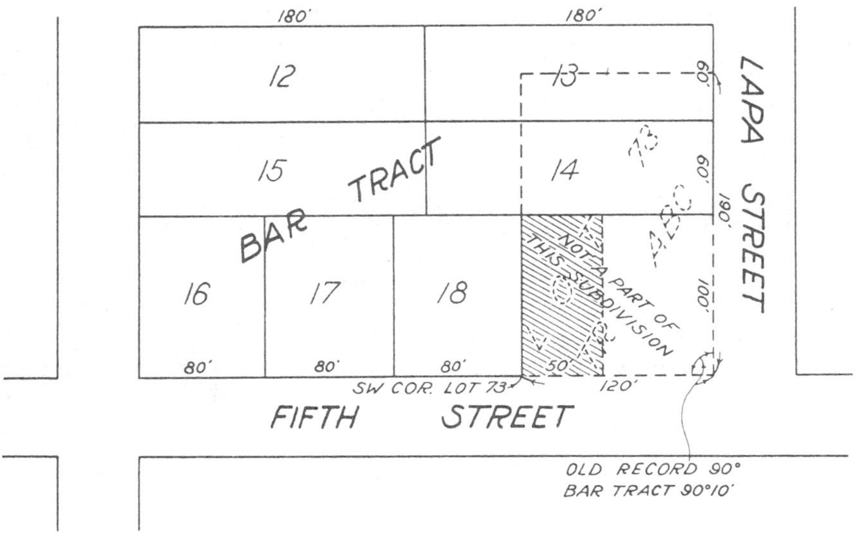

For descriptions and comments covering hatched area, see following #1 to 10.

1. That certain lot, piece or parcel of land lying, being and situate in the City of R, County of Q, State of W, and being a portion of that certain property conveyed to John Doe by that certain indenture dated May 6, 1884, and recorded in Book 146 Page 77 of Deeds, Records of said County, and further shown as a part of Lot 73 of Tract ABC, a map of which was recorded in Book 37 Page 88 of Maps, Records of said County, said parcel of land being more particularly bounded and described as follows, to-wit:

1400 Beginning at a point, said point being in the North line of 5th Street and distant Westerly along said North line of 5th Street 120 feet from the intersection of said North line of 5th Street with the West line of Lapa Street, as said streets are shown on said map of Tract ABC; thence from said point of beginning Northerly and parallel with said West line of Lapa Street a distance of 100 feet to a point; thence Easterly at right angles and parallel with said North line of 5th Street a distance of 50 feet to a point; thence Southerly at right angles and parallel with said West line of Lapa Street 100 feet to a point in said North line of 5th Street; thence Westerly at right angles and along said North line of 5th Street a distance of 50 feet to the point of beginning.

Comment

Too verbose, redundant and repetitive. Unnecessary recitals (as to deed reference). Unnecessary duplication of statements (repeat street names, "point," etc.)

Incorrect assumption of record angles by ignoring matters in latest tract (Bar Tract).

Failure to tie existing boundary (Bar Tract).

Faulty double calls ("right angles and parallel" only one of which is correct).

Excludes attachable street interest. (Recites "along North line of 5th Street.")

Metes and bounds form here is unnecessary.

2. That portion of Lot 73, Tract ABC, in the City of R, County of Q, State of W, as per map recorded in Book 37 Page 88 of Maps in the office of the Recorder of said County, described as follows:

Beginning at the Southwest corner of said lot, in the East line of Lot 18, Bar Tract, as per map recorded in Book 65 Page 22 of said Map Records; thence Northerly along said East line 100 feet to the South line of Lot 14, said Bar Tract; thence Easterly along said South line 50 feet; thence Southerly parallel with said West line 100 feet to the South line of said Lot 73; thence Westerly along said South line 50 feet to the point of beginning.

Comment

A good metes and bounds description with proper ties and a minimum of statement of recitals.

Distances are superfluous except the first "50 feet" which is required to give Easterly limit of the property.

Record assumption which in this instance is valid, that the West line of Lot 73 is same as East line of Lot 18. 1400

3. That portion of Lot 73, Tract ABC, in the City of R, County of Q, State of W, as per map recorded in Book 37 Page 88 of Maps in the office of the Recorder of said County, bounded as follows:

On the South by the South line of said lot; on the West by the East line of Lot 18, Bar Tract, as per map recorded in Book 65 Page 22 of said Map Records; on the North by the South line of Lot 14 of said Bar Tract, and on the East by a line parallel with and distant Easterly 50 feet measured at right angles from said East line of Lot 18.

Comment

A good boundary form description with proper ties and minimum of statement.

4. The West 50 feet of the South 100 feet of Lot 73, Tract ABC, in the City of R, County of Q, State of W, as per map recorded in Book 37 Page 88 of Maps in the office of the Recorder of said County.

Comment

The shortest form of description of the property.

Good only on assumption of record as to boundaries and distances.

Fails to modernize by tie to adjoining tract.

5. The West 50 feet of that portion of Lot 73, Tract ABC, in the City of R, County of Q, State of W, as per map recorded in Book 37 Page 88 of Maps in the office of the Recorder of said County, which lies East of the East line of Lot 18, Bar Tract, as per map recorded in Book 65 Page 22 of said Map Records, and South of the South line of Lot 14, said Bar Tract.

Comment

A good modernized description making proper ties and recitals with a minimum of statement.

6. THE WEST 50 FEET of that portion of Lot 73 in Tract ABC in the City of R, County of Q, State of W, as per map recorded in Book 37, Page 88, of MAPS in the Office of the Recorder of said County included within the area marked "NOT A PART OF THIS SUB-

1400 DIVISION" on the map of the BAR TRACT as per map recorded in Book 65, Page 22, of said MAPS.

NOTE: Reference to part in Lot 73 of original Tract necessary to retain title.

Comment

The shortest form.

Completely modernized and dependent entirely on the data shown on Bar Tract map.

Preferable from a title standpoint to any other description.

7. Tract ABC.

Lot commencing West 70 feet from the corner of Lapa and 5th Streets; thence West 50 feet with uniform depth of 100 feet north part Lot 73.

Comment

Tax description.

Without merit except as a general approximate location.

Usable *only* for tax purposes.

Calls and ties improperly set up and wanting in completeness.

8. West 50 feet of fractional Lot 73, Tract ABC.

Comment

Bond and assessment form description.

Entirely incomplete and inadequate.

Uncertain as to position without an accompanying map.

9. Property at #4617 W. 5th St.

Comment

Common form used by brokers, property owners, post office delivery, contractors, etc.

Location form and usable for general matters only.

Not a competent form for any survey location, or title work.

10. Lot 50 x 100 feet on North side of 5th Street, 70 feet West of Lapa Street.

Comment

Another general location form and usable for such purpose only.

Not sufficient for survey or title purposes.

— Preference of recital —

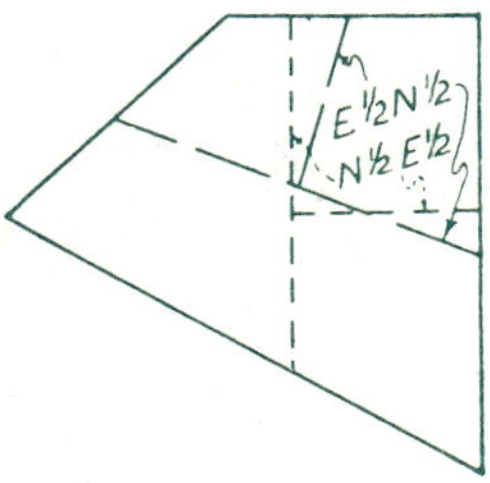

E½ of N½ is same as N½ of E½ in regular parcels but is generally not the same in irregular parcels.

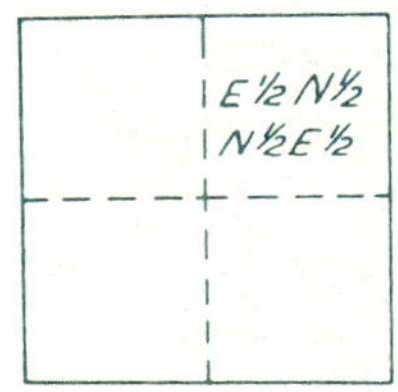

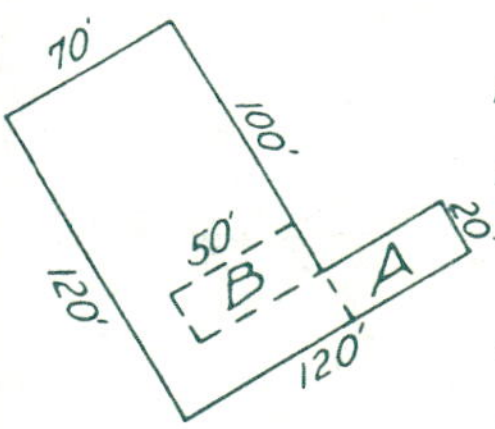

E50 ft of S20 ft covers Parcel A in each case. S20 of E50 ft covers A in each case, but may also cover B.

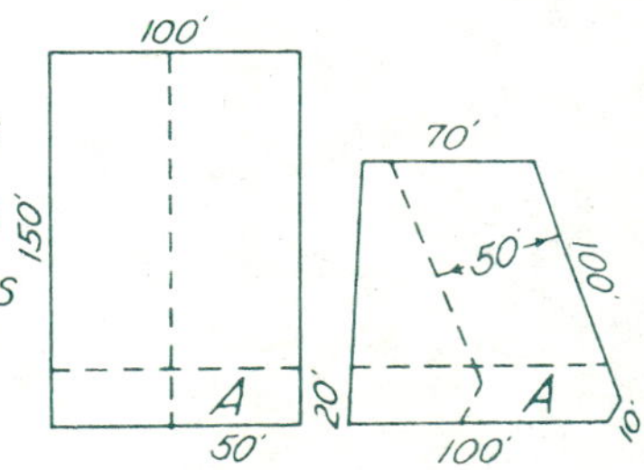

N½ by Areas

1) "N½ of W100' meas along N line of lot;" or
2) "N½ of E100' meas along N line of lot;"

are not the same as

1) "W100' of N½ of lot, meas along N line thereof.
or 2) "E100' of N½ of lot, meas along N line thereof."

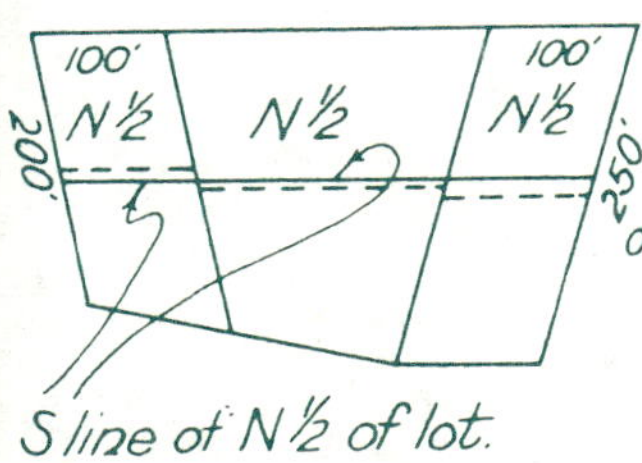

The N½ of the following described ppty:– "Lot except E100 feet meas along N line thereof and except W100 feet meas along N line thereof." is not the same as "N½ of lot exc. E100 feet meas along N line and except W100 feet meas along N line."

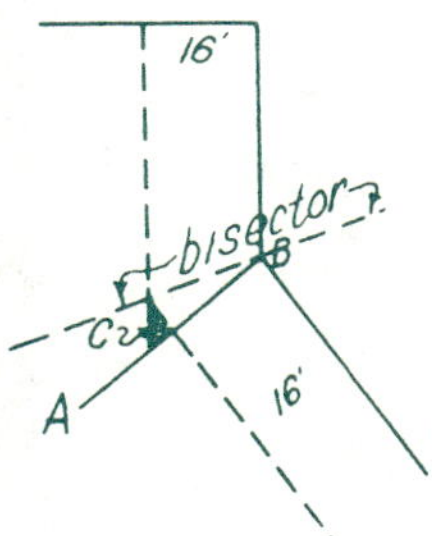

That portion of the E16 feet of lot lying N of AB does not cover C; the E16 feet of that portion of the lot lying N of AB includes C.

1500

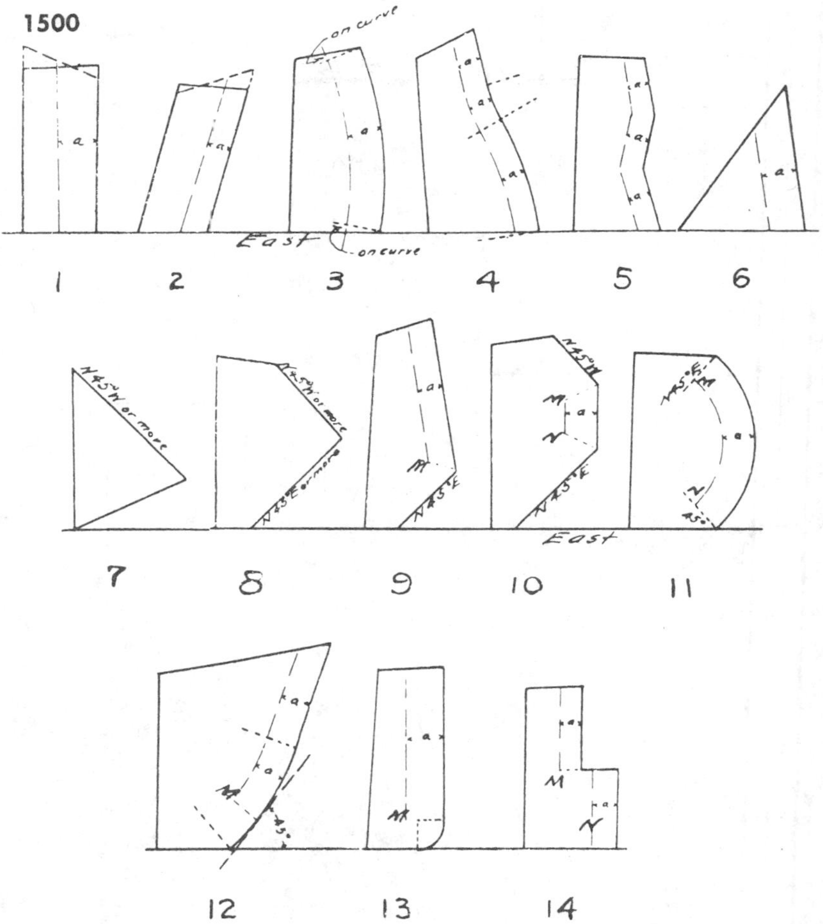

The east (easterly) "a" feet of Lot—

1 to 6 determinate

7 and 8 indeterminate

9-12-13 determinate only above M

10-11 determinate only between M and N

14 indeterminate; may be either M or N, or both

—Fig 1—

1500

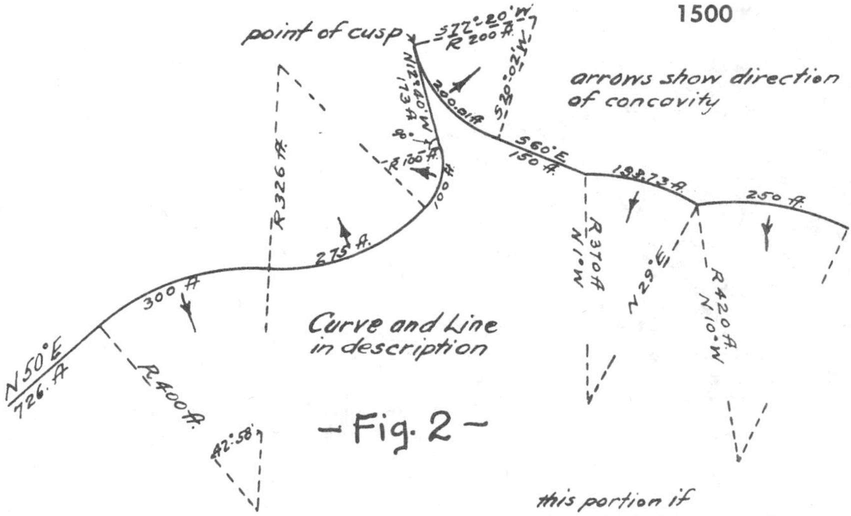

-Fig. 2-

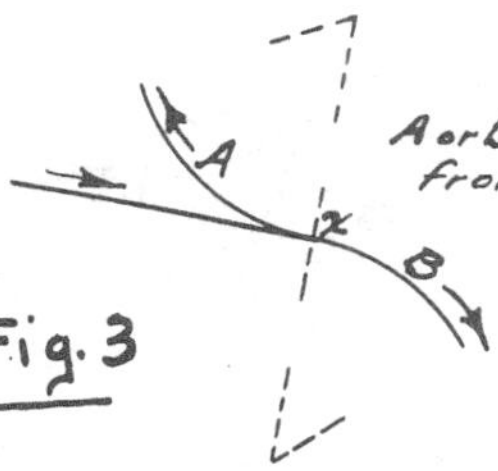

A or B is a "curve to the right," from X

Fig. 3

-Fig. 4-

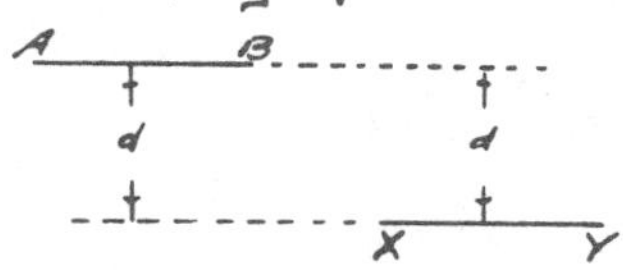

A-B is parallel with X-Y, or either is parallel with the prolongation of the other.

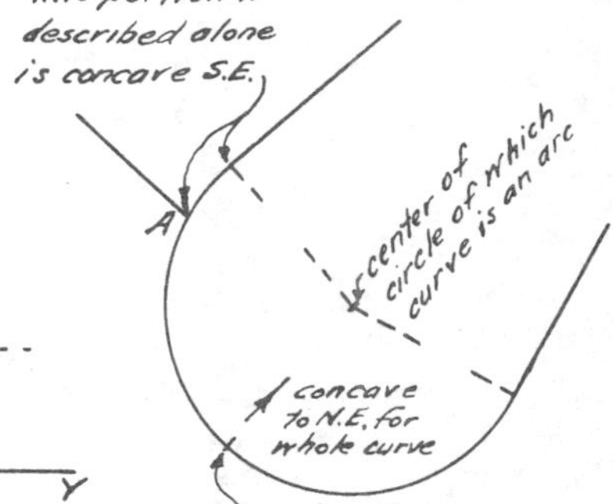

-Fig. 5-

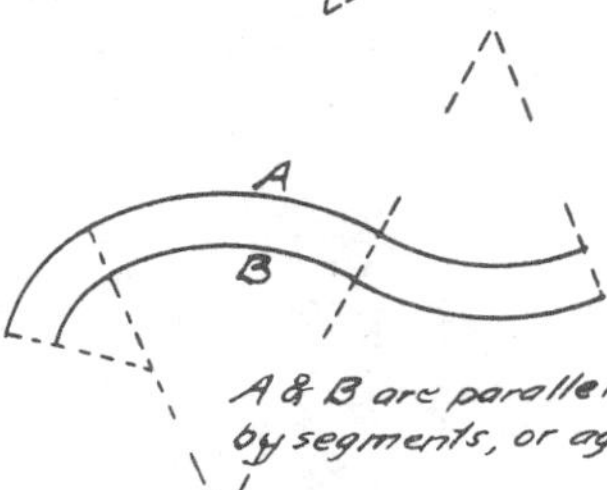

A & B are parallel curves by segments, or aggregate

-Fig. 6-

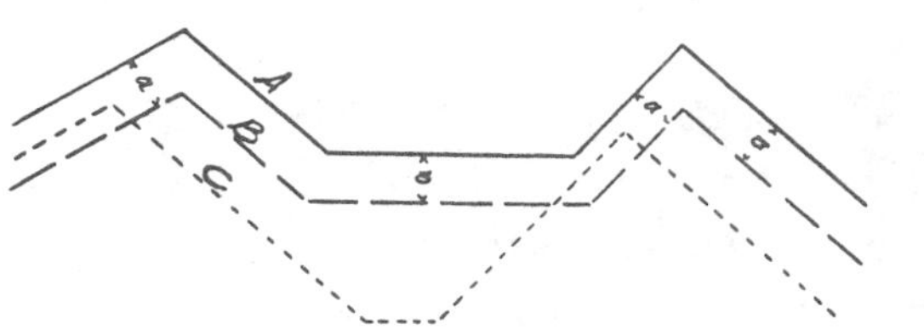

A & B parallel as to segments and aggregate
A & C parallel as to segments but not as to aggregate

-Fig. 8-

-Fig. 7-

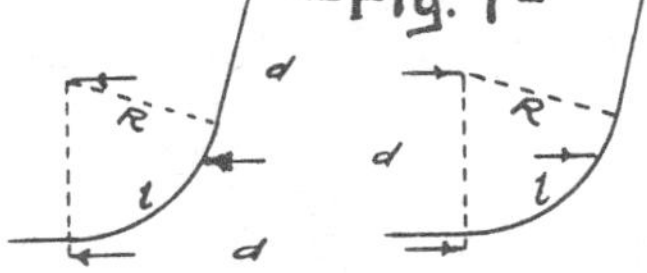

Curves equal but not parallel, although corresponding points are at equal distances along parallel lines.

1500 — Fig 9 —

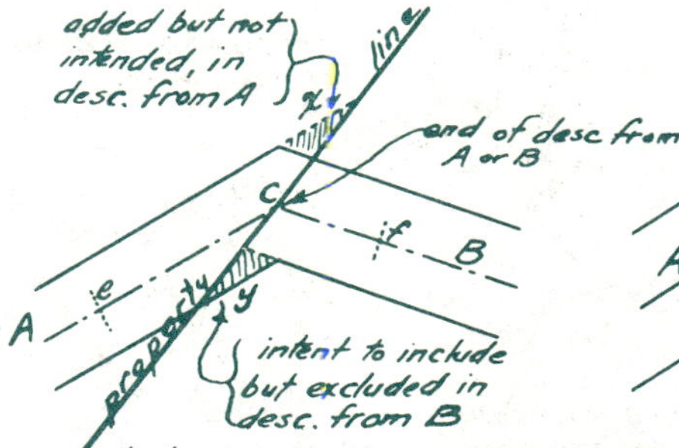

To exclude x, and include y, carry desc. from A through to f, and desc from B through to e.

— Fig 10 —

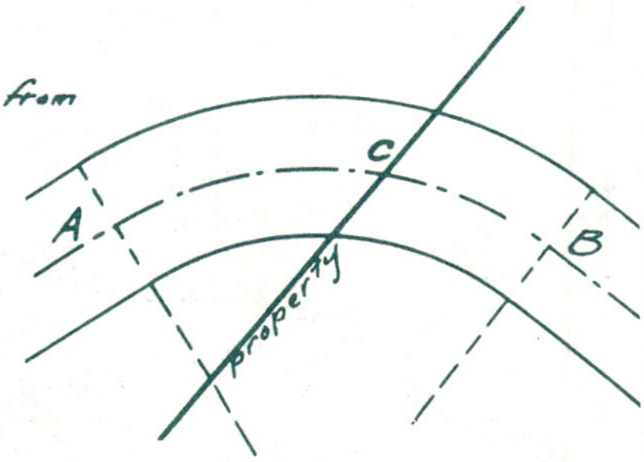

Extension of description beyond C not necessary; extension of curves maintains continuity

— Fig 11 —

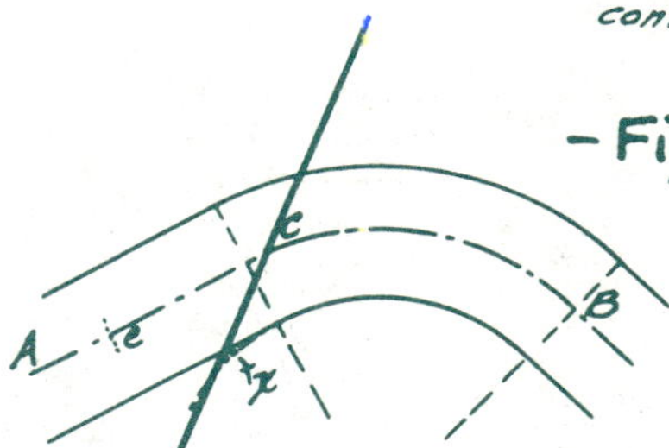

Desc. from A can end at C
Desc from B must continue to e to prevent inclusion of x, not intended.

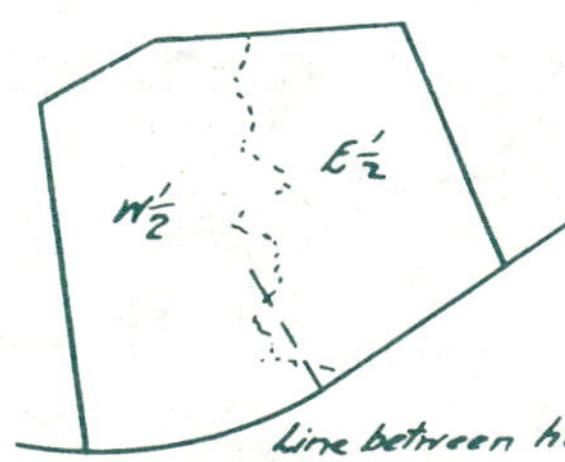

Line between halves indeterminate without qualification.

— Fig 12 —

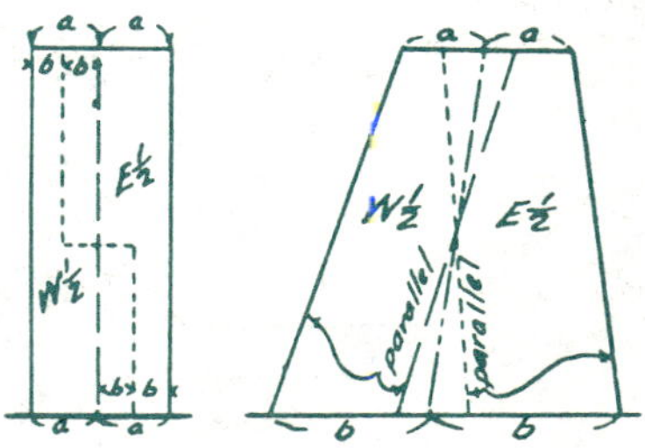

ordinary assumptions

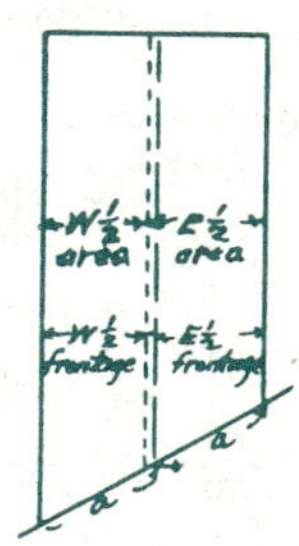

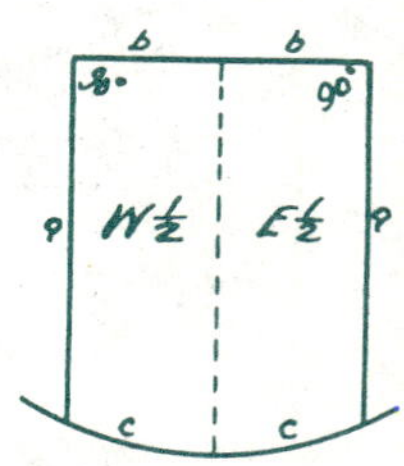

common assumption

Any line dividing an area into equal parts, is a correct line between halves, unless qualified; the half, "frontage", is not half, "area", except in certain cases.

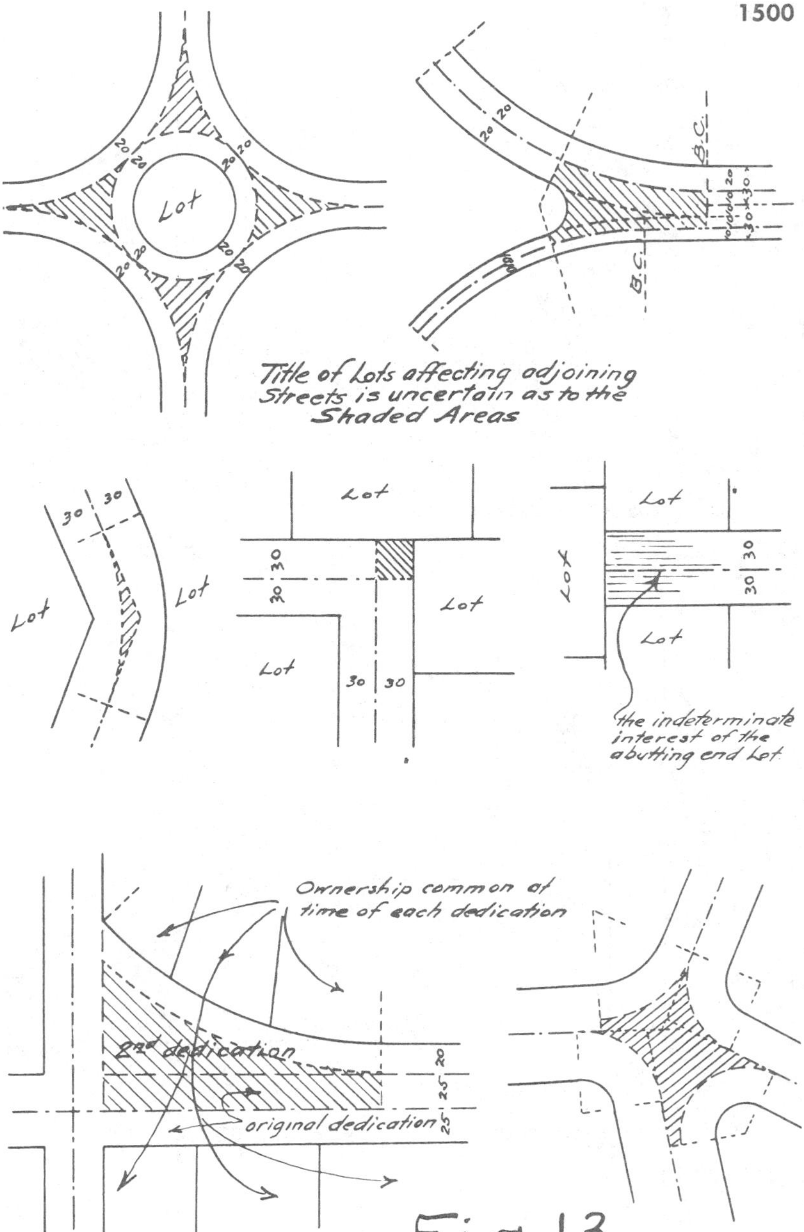
Lot
Title of Lots affecting adjoining Streets is uncertain as to the Shaded Areas
B.C.
Lot
30
Lot
Lot
Lot
Lot
Lot
Lot
Lot
Lot
the indeterminate interest of the abutting end Lot
Ownership common at time of each dedication
2nd dedication
original dedication
20
25
25

Fig. 13

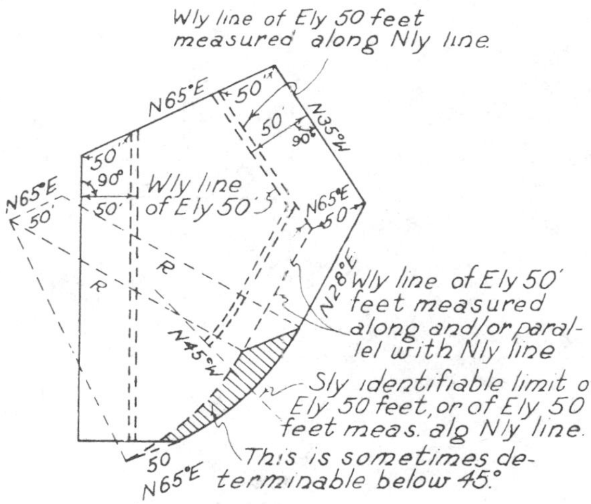

—Fig 14—

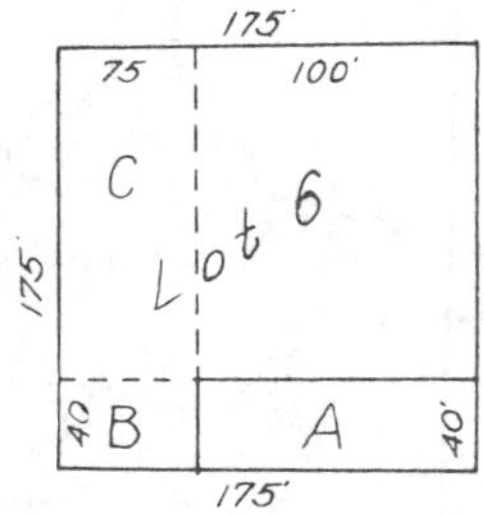

"Lot 6, except the S 40', except the W 75;" intended to except 100'x40' but may have excepted A, B, & C. Describe as "Lot 6 except that portion of the S 40 ft. lying E of the E line of the W 75 feet of said lot".

—Fig 15—

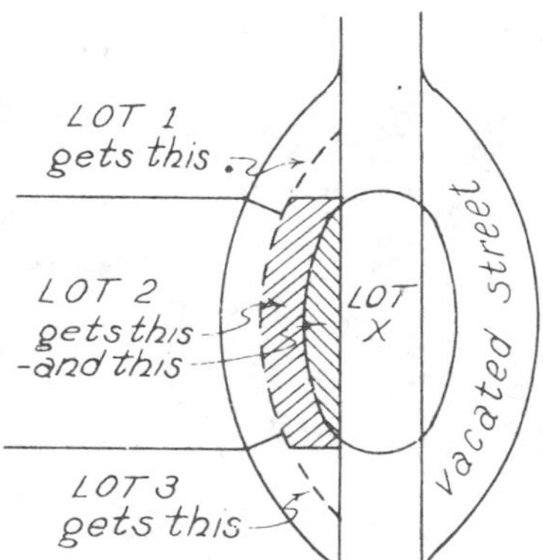

Owner of Lot X deeds to opposite owners on lot lines produced.

—Fig 16—

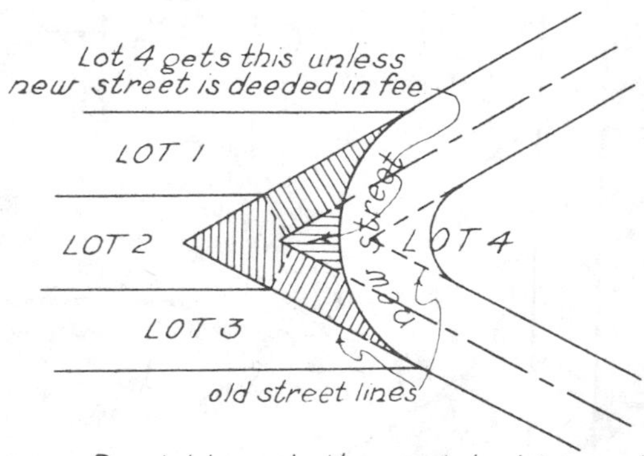

By right angle theory Lots 1, 3, & 4 cut off Lot 2 from new street frontage.

—Fig 17—

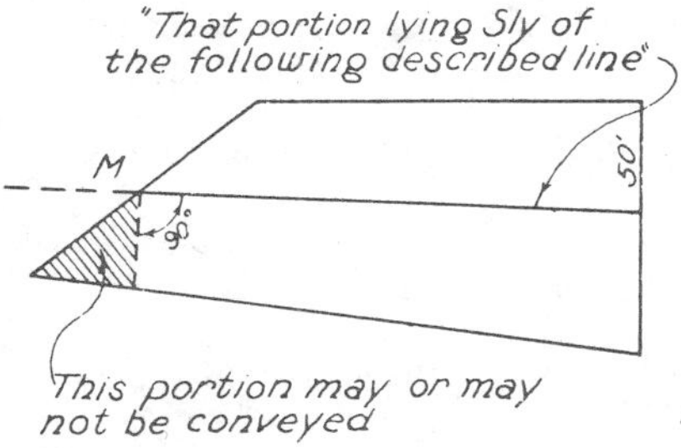

For safety, carry description of line through point "M."

- Fig 18 -

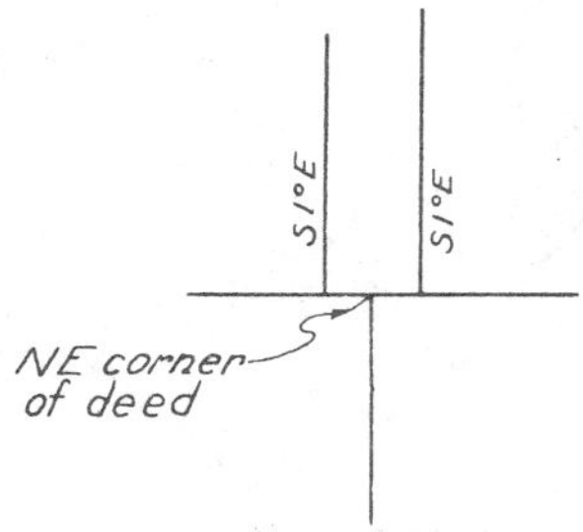

When the preceding course is near corner (one side or the other) but the exact position is indeterminate, describe:-"S 1°E to N line, or its prolongation, of the land described in deed _____; then to the NE corner of the land described in said deed" (No direction or distance.)

Fig 20

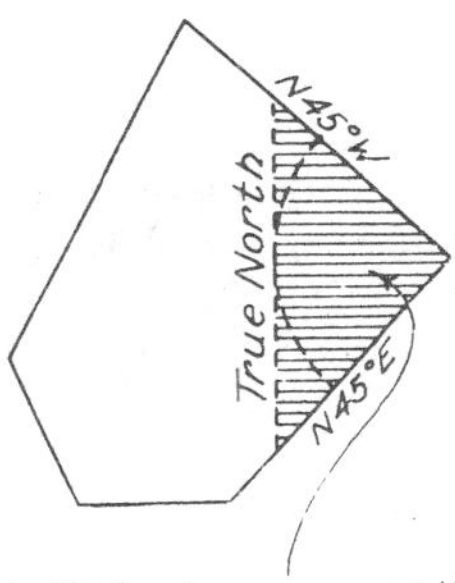

The E 50 feet may possibly be a sector of a circle with a radius of 50 feet or it may be the portion E of due N&S line at 50 feet from apex.

-Fig 19-

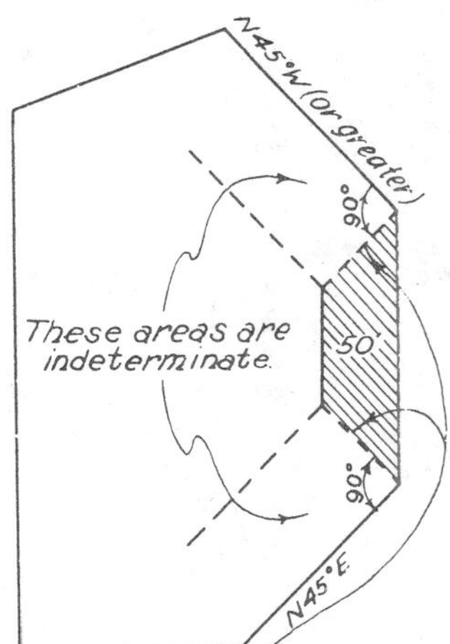

Limits of identifiable area of the E 50 feet.

-Fig 22-

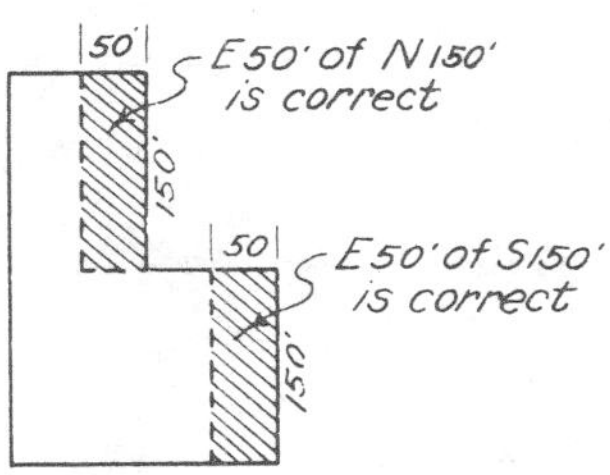

The E 50 feet may be either or both portions.

-Fig 21-

1500

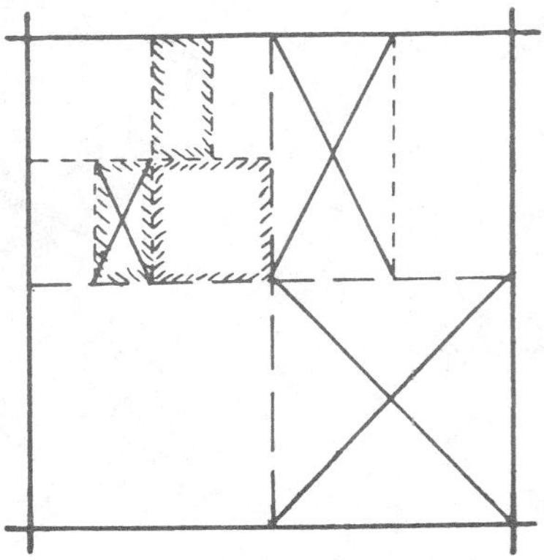

1. "N½ of SE¼ of NE¼ and SW¼ of NE¼ and S½ of NW¼ of NE¼ of Section"

is hatched area and as intended.

2. "N½ of SE¼, and SW¼, and S½ of NW¼, of NE¼ of Section"

is erroneous attempted short form of 1., but is shown by X areas.

Fig 23-

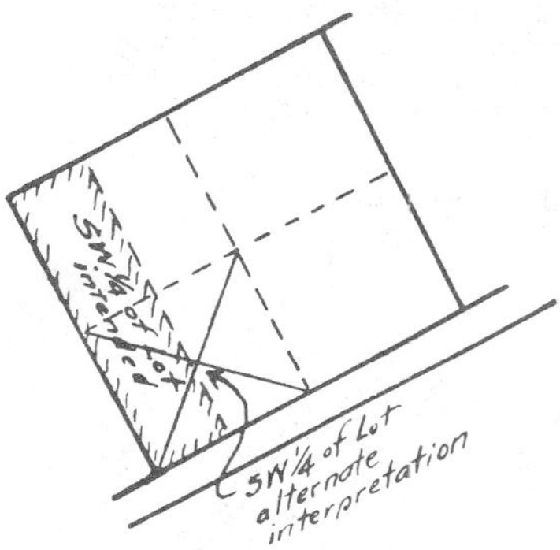

"SW½ of SW½" is proper notation for hatched portion and can not be misconstrued.

-Fig 24-

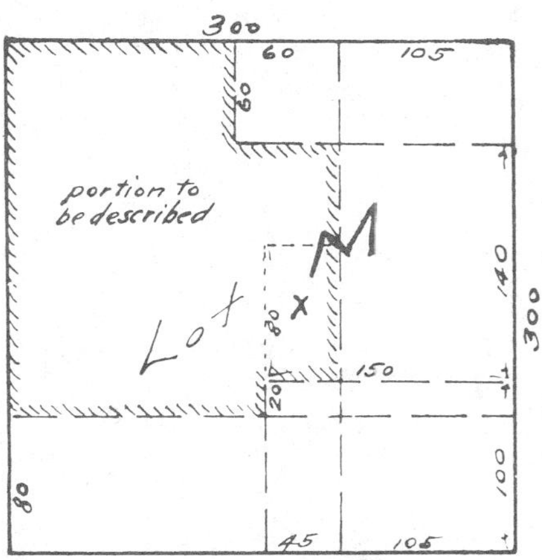

Lot M
Except the E 105 ft.
Also except the N60 ft of the E 165 ft.
Also except the S100 ft of the E 150 ft.
Also except the S80 ft.

The overlapping exceptions present no conflict in the described land.

Lot M
Except the E 105 ft.
Also except the N60 ft of the W60 ft of the E105
Also except the N20 ft of the S100 ft. of the W45 ft of the E150 ft
Also except the S80 ft of said Lot.

Not "except the S80 ft of the remainder; it may also cover "X", not intended

-Fig 25-

ARGENTINA	
34.12 inches	Funk & Wagnalls' Dictionary. Webster's Dictionary.
ARIZONA	
33.00 inches	Official recognition, Court of Private Land Claims.
BOLIVIA	
32.90 inches	Funk & Wagnalls.
BRAZIL	
43.31 inches 43.00 inches	Funk & Wagnalls. Webster.
BRITISH HONDURAS	
32.75 inches	Funk & Wagnalls.
CADIZ	
33.37 inches	Webster.
CALIFORNIA	
(See also Mexico) 33.00 inches—2.75 feet	Hodgman Land Surveying, by sanction of General Land Office for California. Tracy Plane Surveying. Davis & Foote Surveying.
33.004 inches—2.7503 feet based on 1 sq. league equals 4340.277 acres.	Instr. of U.S. Sur. Gen. from original field notes of Geo. Hansen on survey of Pueblo of L.A., 1855.
33.00 inches	Report of U.S. Sur. Gen. of Cal. 11/14/51, as used by "common consent" and sanctioned by GLO 1852.
32.99312 inches—based on 4840 yds. sq. equals 5762.403 vs. sq.	Agreement of Mexican Treaty accompanying above report.
32.953 inches—based on 1 vara equals 837 mm.—or	Above agreement.
32.95269 inches	

1600 CALIFORNIA (continued)

33.312 inches—based on 1 league equals 5000 vs. equals 2.63 miles; 1 sq. league equals 4426.816 acres.	Webster.
33.372 inches—2.781 feet.	Trautwine (the vara "by law") (Note: There is no statute in California establishing the length of the vara.)
33.372 inches—2.781 feet— (100. v equals 4.21 ch.) (210.68 ch equals 1 league equals 5000 v)	L.A. Co., M.R. 53-59. Pat. 2-38.
33.372 inches—2.781 feet	L.A. Co. Pat. 2-38 & 2-444. L.A. City Lands (1 league each way from Plaza equals 5000 v and by Pat. equals 210.68 ch.); also M.R. 1-401; also course #1 of bdy of Ro Ex Mission de San Fernando equals 5000 v equals 210.68 ch., by Pat. 1-606.
from the last above three: 1 ch. equals 23.7327 v. 1 v. equals 0.0421 ch. equals 4.2136 lks. 1 sq. league equals 25,000.000 sq. v. equals 4438.606 acres. 5000 v. equals 13904.88 ft. equals 2.633 miles.	
The proper value of the vara, by record, to be used in L.A. County as derived above, is 2.781 ft., or 0.0421 ch.	
32.99312 inches—2,74943 feet—4.1658 lks.	Public Domain in the Southwest, War Dept. Education Manual EM-908.
33.3333 inches — based on 1 league as equaling 4428.4 acres.	U.S. vs. Perot; 98 US 428 (1878).

		1600
Central America		
33.87 inches	Webster.	
Chile		
32.91 inches	Funk & Wagnalls.	
33.37 inches	Webster.	
Colombia		
31.50 inches	Funk & Wagnalls.	
	Webster.	
Costa Rica		
32.88 inches	Funk & Wagnalls.	
Cuba		
33.38 inches	Funk & Wagnalls.	
	Webster.	
Curaçao		
33.37 inches	Funk & Wagnalls.	
Florida		
33.372 inches	U.S. War Dept. Educational Manual EM-908.	
	"Castilian" vara.	
	Mitchel vs US; 15 Peters 51,57 (1841).	
Honduras		
33.03 inches	Funk & Wagnalls.	
33.00 inches	Webster.	
New Mexico		
33.00 inches	Surveys of private claims; official recognition by Court of Private Land Claims.	
Mexico		
conversions by relation to millimeters based on 1 meter equals 39.37 inches, or 1 yard equals 36 inches equals 3600/3937 meters. 1 meter equals 1000 millimeters.	Merriman's Civil Engr. Pocketbook.	
33.03773 inches—based on 1 vara equals 839.16 mm.	Humboldt (1803).	

1600 MEXICO (continued)

32.95269 inches—based on 1 vara equals 837 mm.	Agreement between Mexican Republic and Bond Holders 9/15/1837 from report. Commr. GLO 1854.
32.99312 inches	Same.
32.99206 inches—based on 1 vara equals 838 mm.	Orbezogo (about 1844). Also measured interval between marks on old standard rod from Arizpe, Sonora, Mexico 838 mm.
32.96568 inches—1 vara equals 837.33 mm.	Bustamente 1851.
32.96753 inches—1 vara equals 837.377 mm.	U.S. Coast Survey 1850.
32.9682 inches	Standard from Mexico at close of Mexican War, in office of U.S. Coast Survey. U.S. vs. Perot 98. US 428 (1878).
33.00 inches	U.S. vs. Perot. Report Commr. GLO, 1854 as by common consent. Webster.
32.994 inches — based on 1 sq. vara equals 0.84 sq. yd.	Webster.
32.99 inches	Funk & Wagnalls.
32.992 inches	Merriman C.E. Pocketbook. Tracy Plane Surveying.
32.993 inches	Davis & Foote Surveying.
33.38676 inches or 0.92741 yard.	J. H. Alexander Dictionary. Weights and Measures.
32.808 inches—based on geometrical ft. equals 10.936 in., and 1 vara equals 3 pies geometricos.	Haggard Handbook for Translators of Spanish Historical Documents.

Paraguay	
33.00 inches	Funk & Wagnalls.
34.12 inches	Webster.
Peru	
32.91 inches	Funk & Wagnalls.
33.37 inches	Webster.
Philippine Islands	
33.38 inches	Webster.
Portugal	
43.28 inches	Funk & Wagnalls.
43.00 inches	Webster.
Spain	
33.36 inches	Funk & Wagnalls.
33.38 inches	Webster.
32.875 inches – based on 1 Castilian ft. equals 278.3332 mm.	Bruhn Log. Tables & Trautwine.
32.913 inches	Merriman C.E. Pocketbook.
Texas	
33.40 inches–based on geometrical ft. equals 11.13333 inches, and 3 geo. ft. equals 1 vara (Spanish Bar).	Seth Ingram, Colonial Surveyor, in document. 8/18/1825.
39.60 inches – assuming 10 bars (varas) equals 50 links (33 ft.).	Seth Ingram's chain used for measurement.
33.3333 inches–or 36 vs. equals 100 feet.	Hodgman Land Surveying; Established by Texas Law 1919.
Venezuela	
33.38 inches	Funk & Wagnalls.

The variance and uncertainty of the value of the vara as disclosed by the foregoing table requires a definite determination of the value used in any particular instance, when a relocation of a line (whose original length was expressed in varas) is to be made.

The established length, by law (Texas), or by the courts in specific suits (U.S. vs. Perot, for example, etc.), must be held if the problem involves land governed thereby, but a relocation to be as exact as possible, must depend on the discovered value nearest that used in the original location. The principle of "common consent" is valueless in relocation, if other considerations or values intervene; a good example is indicated in Los Angeles County. The length of the vara here is established by no less than 5 different instruments of record as 33.372 inches, in spite of the reports of various agencies that "common consent" shows 33.00 inches.

Before making any survey or computation affecting a parcel of land whose original dimensions are given in varas, it is incumbent on the surveyor to discover, as nearly as possible, the true value of the vara used in the instant parcel dimensions, and convert to present day lengths on that basis. Any maps or documents referring to such original measurements should state the source and conversion factor used to fix the modernized measurements.

Table of Vara Values
Compiled, with memorandum, by
WM. C. WATTLES
L.S. 1071, California

INDEX